DC Theory

NJATC

DC Theory

NJATC

THOMSON

DELMAR LEARNING   Australia   Canada   Mexico   Singapore   Spain   United Kingdom   United States

THOMSON

DELMAR LEARNING

DC Theory
NJATC

Vice President, Technology and Trades SBU:
Alar Elken

Executive Director, Professional Business Unit:
Gregory L. Clayton

Product Development Manager:
Patrick Kane

Development Editor:
Angie Davis

Channel Manager:
Beth A. Lutz

Marketing Specialist:
Brian McGrath

Production Director:
Mary Ellen Black

Production Manager:
Larry Main

Senior Project Editor:
Christopher Chien

Art/Design Coordinator:
Francis Hogan

COPYRIGHT © 2004 by National Joint Apprenticeship Training Committee

Printed in the United States of America
1 2 3 4 5 XX 06 05 04 03

For more information contact
Delmar Learning
Executive Woods
5 Maxwell Drive, PO Box 8007,
Clifton Park, NY 12065-8007
Or find us on the World Wide Web at
www.delmarlearning.com

For permission to use material from the text or product, contact us by
Tel.   (800) 730-2214
Fax   (800) 730-2215
www.thomsonrights.com

Library of Congress Cataloging-in-Publication Data:

NJATC.
   DC Theory / NJATC.
      p. cm.
      Includes bibliographical references and index.
      ISBN 1-40185-686-1
   1. Electric engineering. 2. Electric circuits—Direct current. 3. Electricity. I. Title. TK146 .C19 2004 621.319'12—dc22

                    2003020321

NOTICE TO THE READER

Contents

CHAPTER 2
Mathematical Concepts for Solving Electrical Problems

26

CHAPTER 3
Using Ohm's Law and Associated Electrical Units 52

CHAPTER 4
The Properties of Power in an Electrical Circuit

68

CHAPTER 6
How Current Reacts in a DC Circuit   104

CHAPTER 7
How Voltage Functions in a DC Series Circuit   118

CHAPTER 8
How Voltage Dividers Work in a DC Series Circuit — 132

CHAPTER 9
How to Calculate Power in a DC Series Circuit   138

PART 3
DC PARALLEL CIRCUITS   147

CHAPTER 10
How Voltage Functions in a DC Parallel Circuit 148

CHAPTER 11
Understanding Resistance in a DC Parallel Circuit   156

CHAPTER 12
How Current Reacts in a DC Parallel Circuit   166

CHAPTER 13
How Current Dividers Work in a DC Parallel Circuit   174

CHAPTER 14
How to Calculate Power in a DC Parallel Circuit 180

PART 4
DC COMBINATION CIRCUITS 185

CHAPTER 15
Understanding Resistance in Combination Circuits 186

CHAPTER **16**

How Current Reacts in DC Combination Circuits   194

CHAPTER **17**

How Voltage Functions in DC Combination Circuits   202

CHAPTER 18
How to Calculate Power in DC Combination Circuits
212

PART 5
MAGNETISM AND GENERATORS
221

CHAPTER 19
Understanding the Principles of Magnetism and Electromagnetism
222

CHAPTER 20
How Electrical Generators Work   **234**

PART 6
DC CIRCUIT ANALYSIS TOOLS   245

CHAPTER 21
Understanding Voltage Polarity and Voltage Drop   246

CHAPTER 22
Applying the Principle of Superposition   258

CHAPTER 23
Using Kirchhoff's Laws to Solve DC Circuits   268

CHAPTER 24
Thevenin's and Norton's Theorems   278

Index   287

Preface

The National Joint Apprenticeship and Training Committee (NJATC) is the training arm of the International Brotherhood of Electrical Workers and National Electrical Contractors Association. Established in 1941, the NJATC has developed uniform standards that are used nationwide to train thousands of qualified men and women for demanding and rewarding careers in the electrical and telecommunications industries. To enhance the effectiveness of this mission the NJATC has partnered with Delmar Learning to deliver the very finest in training materials for the electrical profession.

Knowledge of fundamentals is critical to the success of a modern electrical technologist. Every project, every piece of knowledge, and every new task will be based on all of the experience and information that you get as you progress through your career. This book contains much of the material that will form the foundation of your electrical knowledge.

Although based on direct current (DC) circuits, the methods and techniques that you will learn in this book will apply to AC circuits as well. This means that you must study carefully, commit the DC principles to memory, and understand not just the methods but also the underlying theories and concepts.

To help in organizing the learning process, this book is divided into six parts. Each part contains chapters that progress in a rational, well-paced schedule that allows you to become thoroughly familiar with one set of principles before you move on and apply your new knowledge to learning more advanced topics.

PART 1—FUNDAMENTALS OF ELECTRICITY AND DC CIRCUITS

In Chapter 1 you will learn the physical principles that underlie electrical technology. Starting with atomic structure the chapter progresses to a thorough discussion of the electrical properties of materials and how they are categorized into conductors, semiconductors, and insulators. The chapter covers aspects of electrical theory including molecules, ions, electrolytes, and a variety of other knowledge that will provide an excellent foundation. The chapter concludes with an introduction to the ways in which electricity is generated with special emphasis on the types of "real world" applications that the practicing electrical technologist will encounter in his or her career.

Chapter 2 provides a review of mathematical and algebraic operations that are fundamental for the day-to-day work of the electrician. The chapter begins with an introduction of the prefix multipliers such as milli, kilo, mega, and micro that are necessary to understanding electrical systems. The chapter then covers the creation, analysis, and solution of mathematical equations that are encountered in electrical work. This chapter is provided as a review; however, we strongly recommend that the student work through the examples to be certain that the knowledge picked up in algebra courses is as fresh as it should be.

Chapter 3 introduces the single most important formula in electricity—Ohm's law. The usual method of using Ohm's law is applied; however, the chapter does not stop where many other texts do. The nature of resistance in conductors is discussed and formulas that allow the user to calculate resistances including the effects of material, size, and heat are presented.

Chapter 4 extends the material from Chapter 3 to include more in-depth information on the use of Ohm's law as it relates to voltage, current, resistance, and power. This chapter also provides more detail for the definitions of some of the basic units and shows how to calculate various values for actual equipment.

PART 2—DC SERIES CIRCUITS

Chapters 5 through 9 cover the theory and practical applications of electrical components connected in series. Chapters 5, 6, 7, and 9 have detailed explanations on calculating and measuring resistance, voltage, current, and power along with numerous examples and problems allowing the student to practice the procedures. The use and calculation of voltages in a voltage divider are presented in detail in Chapter 8.

PART 3—DC PARALLEL CIRCUITS

This part builds on the previous parts by teaching the manipulation and understanding of electrical elements connected in parallel. Chapters 10, 11, 12, and 14 include detailed explanations on calculating and measuring resistance, voltage, current, and power. As in Part 2, numerous examples and problems that allow the student to practice the procedures are included. The use and calculation of currents in a current divider are presented in detail in Chapter 13.

PART 4—DC COMBINATION CIRCUITS

Chapters 15 through 18 complete the fundamental introductions begun in Parts 2 and 3. Here the student learns and practices the analysis of DC circuits that are made of a combination of series and parallel branches. Practice problems again allow the student to exercise the newly learned skills.

PART 5—MAGNETISM AND GENERATORS

Chapters 19 and 20 introduce the student to the principles of magnetism and electromagnetism. The nature of magnetism, its source, how it is analyzed, how it is created, and how electricity is created using magnetism are explained in Chapter 19.

Chapter 20 combines the electrical and magnetic knowledge previously covered to introduce the student to the real world of electrical generators. The basic concepts of generation are provided, allowing future courses to provide the more in-depth coverage that will be required by the practicing electrician.

PART 6—DC CIRCUIT ANALYSIS TOOLS

The final chapters present three advanced techniques that the electrician will find invaluable in analyzing and understanding the more complex circuits he or she will encounter on the job. Chapter 21 introduces the concepts of voltage polarity and voltage drop as well as providing early instruction on the selection and application of conductors for a given circuit. Chapters 22, 23, and 24 teach the user of superposition, Kirchhoff's laws, and Thevenin's and Norton's theorems. These final chapters complete the initial introduction of the basic electrical principles to the student.

Taken together, these 24 chapters, when accompanied by classroom presentations and/or organized self-study, embody a comprehensive introduction to basic electrical theory.

The NJATC can provide a complete line of electrical & telecommunication training materials, including CBT programs and courses. Visit the NJATC online at njatc.org to review the finest electrical training curriculum the industry has to offer. The subject of Direct Current is both interesting and essential for the electrical and electronic student. Take the time to progress through the DC text material, perform the calculations and review the chapter objectives before moving forward to the next section. Your understanding of DC Theory will provide all of the essentials to move to the next level of expertise in the electrical and electronic fields. Should you decide on a career in the electrical industry, the International Brotherhood of Electrical Workers and the National Electrical Contractors Association (IBEW-NECA) training programs provide the finest electrical apprenticeship programs in the industry. If you are accepted into one of their local apprenticeship programs you'll be trained for one of four career specialties, journeyman lineman, residential wireman, journeyman wireman or VDV installer/technician. Most importantly, you'll be paid while you learn. To learn more visit http://www.njatc.org.

NJATC ACKNOWLEDGEMENTS
Principal Writer

Stan Klein, NJATC Staff

Contributing Writer

Jim Paladino, Training Director, Omaha, NE

ADDITIONAL ACKNOWLEDGEMENTS

This material is continually reviewed and evaluated by Training Directors who are also members of the NJATC Inside Education Committee. The invaluable input provided by these individuals allows for the development of instructional material that is of the absolute highest quality. At the time of this printing the Education Committee was comprised of the following members:

INSIDE EDUCATION COMMITTEE

Dennis Anthony–Phoenix, AZ; John Biondi–Vineland, NJ; Dan Campbell–Tangent, OR; Peter Dulcich–Syracuse, NY; John Gray–San Antonio, TX; Gary Hunziker–Sacramento, CA; Dave Kingery–Salt Lake City, UT; Bill Leigers–Richmond, VA; Bud McDannel–West Frankfort, IL; Bill McGinnis–Wichita, KS; Jerry Melson–Bakersfield, CA; Tom Minder–Fairbanks, AK; Bill Newlin–Dayton, OH; Jim Paladino–Omaha, NE; Dan Sellers–Collegeville, PA; and Jim Sullivan–Winter Park, FL.

OUTSIDE EDUCATION COMMITTEE

Charley Young–Lawrence, KS; S. K. Pelch–Sandy, UT; Armando Mendez–Riverside, CA; Steve Uhl–Limerick, PA; Bill Stone–Portland, OR; Don Jamison–Indianola, IA; Howard Miller–Medway, OH and Virgil Melton–Atlanta, GA.

PUBLISHER ACKNOWLEDGEMENTS

John Cadick, P.E., Contributor

A registered professional engineer, John Cadick has specialized for almost four decades in electrical engineering, training, and management. In 1986 he founded Cadick Professional Services (forerunner to the present-day Cadick Corporation), a consulting firm in Garland, Texas. His firm specializes in electrical engineering and training, working extensively in the areas of power system design and engineering studies, condition-based maintenance programs, and electrical safety. Prior to the creation of Cadick Corporation, John held a number of technical and managerial positions with electric utilities, electrical testing firms, and consulting firms. Mr. Cadick is a widely published author of numerous articles and technical papers. He is the author of the *Electrical Safety Handbook* as well as *Cables and Wiring*. His expertise in electrical engineering as well as electrical maintenance and testing coupled with his extensive experience in the electrical power industry makes Mr. Cadick a highly respected and sought after consultant in the industry.

Monica Ohlinger

The publisher would like to thank Monica Ohlinger of Ohlinger Publishing Services for her diligent work in development on the text.

PART

1

FUNDAMENTALS OF ELECTRICITY AND DC CIRCUITS

CHAPTER 1

Sources and Effects of Electricity—
The Electron Theory

CHAPTER 2

Mathematical Concepts for Solving
Electrical Problems

CHAPTER 3

Using Ohm's Law and Associated
Electrical Units

CHAPTER 4

The Properties of Power in
an Electrical Circuit

Sources and Effects of Electricity—The Electron Theory

■ OUTLINE

■ OVERVIEW

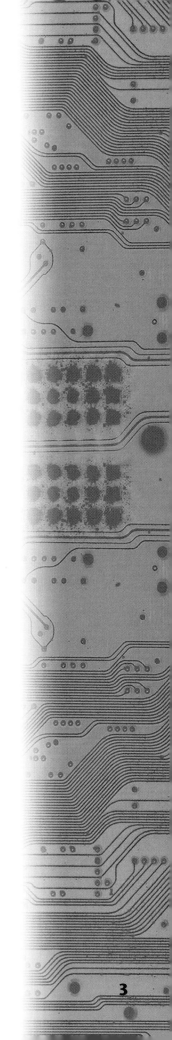

Very early in the twentieth century a sleeping student was awakened by his professor's question, "James, what is electricity?"

Sleepy and confused, the student tried vainly to recover. "Er-r-r, well, um-m-m, I did know, sir. But I forgot."

"What?" cried the smiling professor. "You are the only person in history to know what electricity is—and you forgot!?!"

Intensive research and investigation in the eighteenth and nineteenth centuries had shown much about the behavior of electricity but little about its roots. Consequently, the unfortunate student's statement that "I did know" pre-dated anyone knowing the true nature of electricity.

The real understanding of electrical fundamentals began with J. J. Thompson's 1897 discovery of the electron. However, a comprehensive knowledge of electricity was slow in developing, requiring many subsequent twentieth-century revelations about the fundamental structure of the atom.

In this chapter you will learn the so-called *fundamentals* of electrical theory. The fundamentals include the following topics:

- What is electricity?
- Where does electricity come from?
- What type of objects or materials exhibit electrical behavior?
- What sources can be used to create and manipulate electricity?

This chapter presents the foundation for all the subsequent chapters. Study it well and commit the principles to memory. The principles will serve you well as you progress through your training.

■ OBJECTIVES

After completing this chapter, you should be able to:

1. Describe the basic structure of the atom.
2. Name the three main particles that are part of all but the simplest atom.
3. Describe the electrical characteristics of an atom.
4. Describe the relationship between the valence electrons and electron movement (current flow).
5. Describe different means of producing electrical current.
6. Explain the effects of electrical current.

■ GLOSSARY

Atom   The smallest piece of an element that has all the characteristics of that element.

Battery   A series and/or parallel combination of cells. A group of cells connected in series (more voltage) or parallel (more current).

Cell   A single chemical structure composed of an electrolytic solution (such as sulfuric acid) and two different metallic electrodes (lead and lead peroxide).

Compound   A material made from the chemical combination of two or more elements.

Conductor   A material that easily passes electrical current. Examples include silver, copper, and aluminum.

Electricity   A class of phenomena that results from the interaction of objects that exhibit a charge (electrons and protons). In its static form, electricity exhibits many similarities to another naturally occurring force—magnetism.

Electrolyte   Any material that will dissolve into ions when immersed in a liquid. The liquid thus becomes an electrical conductor.

Electron   One of the three main components of an atom. The electron is a fundamental particle, and by definition it has a negative electrical charge (from the Greek word *elektron,* meaning "to be like amber").

Element   The simplest form of matter. There are over 103 known elements, 92 of which occur naturally. All matter is made from chemical combinations of elements.

Gluon   The particle that mediates or transmits the strong nuclear force between quarks. The fundamental particle that is responsible for the strong nuclear force.

Hadrons   Large, nonfundamental particles. Protons and neutrons are examples.

Insulator   A material that does not pass electrical current easily. Examples include rubber, plastic, and mica.

Ion   An atom or molecule that has gained or lost one or more electrons. A positive ion has lost electrons, and a negative ion has gained one or more electrons.

Isotope   One of two or more atoms with the same number of protons but different number of neutrons.

Leptons   Small, fundamental particles. The electron is an example of a lepton.

Magnetism   A naturally occurring force that attracts certain types of ferrous materials.

Matter   The material from which all known physical objects are made.

Molecule   The chemical combination of two or more atoms. The smallest particle of a compound that has the same chemical characteristics of the compound.

Neutron   One of the three main components of an atom. The neutron has no electrical charge and is therefore classed as electrically neutral. The neutron has been shown to be composed of even smaller particles called *quarks.*

Piezoelectric effect   Electricity being created by stress or pressure in a material—especially a crystalline material.

Primary cell   A cell that cannot be recharged after it has depleted all its stored chemical energy in the form of electricity.

Proton   One of the three main components of an atom. By definition, the proton has a positive electrical charge. The proton has been shown to be composed of even smaller particles called *quarks.*

Quark   One of the fundamental particles of matter. There are six different types of quarks that are assembled in different combinations to create larger particles, such as protons and neutrons.

Residual strong interaction   The force that holds the nucleus of an atom together. The root cause is the attraction between quarks in each of the individual hadrons.

Secondary cell   A cell whose chemical energy can be restored by forcing electrical energy into it.

Semiconductor   A material that falls between conductors and insulators in terms of electrical conductivity.

Static electricity   An electrical charge that is stationary or nonmoving. Sometimes called *triboelectricity.*

Strong nuclear force   The force that holds the quarks together to make up neutrons and protons. The residual strong nuclear force is also responsible for holding the protons and neutrons together in the nucleus despite the electrical repulsion trying to force the protons apart.

Thermocouple   Two dissimilar metals that create an electrical potential when heated.

Thermoelectricity   Electricity created by heat.

Triboelectricity   An electrical charge created by rubbing two materials together. Sometimes called *static electricity.*

Valence electrons   The electrons that make up the valence shell. Valence electrons are free to participate in current flow.

Valence shell   The outermost shell of electrons in an atom. Also called the *valence ring.*

■ INTRODUCTION

Electricity is an invisible force that can produce heat, motion, light, and many other physical effects. This invisible force provides power for lighting, radios, motors, heating and cooling of buildings, and many other applications. The common link among all these applications is the electrical charge. All the materials we know—gases, liquids, and solids—contain two basic particles of electric charge: the electron and the proton. The electron has an electric charge with a negative polarity. The proton has an electric charge with a positive polarity.

1.1 Historical Background

Electricity, electron, and *electronics* are English words that come from a word with a Greek background—*elektron.* The literal English translation of this word is "to be like amber." Over 2,500 years ago, the Greeks found that by rubbing amber with other materials, it became charged with this invisible force and could attract bird feathers, hair, cloth, and other materials. Later, in the 1600s, William Gilbert found that along with amber, other materials could also be charged with this invisible force. He categorized those materials that could be charged as *electriks* and those materials that could not be charged as *nonelectriks.* About a hundred years later, in 1733, Charles DuFay discovered that some charged materials would attract other objects and that other charged objects would repel different objects. Benjamin Franklin suggested the "convention" that two kinds of charges existed—"positive" (+) and "negative" (−)—and that "like" charges repel (positive from positive and negative from negative) and "unlike" charges attract (positive to negative). An example of how these charged materials were classified and how they react to each other is shown in Figure 1–1.

FIGURE 1–1 Source and characteristics of charged materials.

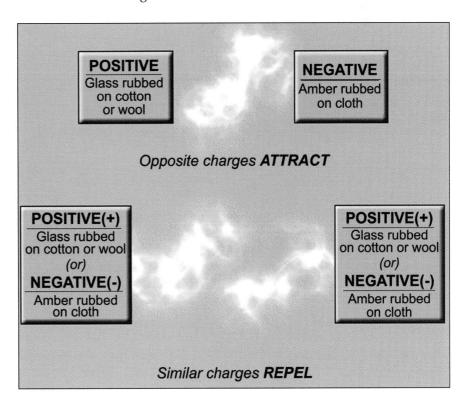

POSITIVE
Glass rubbed
on cotton
or wool

NEGATIVE
Amber rubbed
on cloth

*Opposite charges **ATTRACT***

POSITIVE(+)
Glass rubbed
on cotton or wool
(or)
NEGATIVE(-)
Amber rubbed
on cloth

POSITIVE(+)
Glass rubbed
on cotton or wool
(or)
NEGATIVE(-)
Amber rubbed
on cloth

*Similar charges **REPEL***

FIGURE 1–2   Periodic table of the elements.

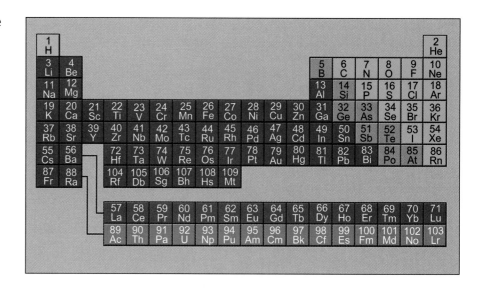

ATOMS AND ELECTRICITY

1.2 Matter and Atoms

Look around you. Concentrate on a chair, a table, or anything that is not living. These things are objects. Now (pretend) tear or break the object apart (chop it into firewood or tear it into pieces). The object no longer exists in its original form, the form in which you first saw it. What is left is the material or matter that was formed or composed into some shape to make up the original object. All things in the universe are "assembled" or comprised of matter. **Matter** is the "stuff" that makes up the universe. Matter can exist in the universe in one of three forms: a solid, a liquid, or a gas. Matter can exist in all three forms but not at the same time. A good example of this is water. Water is naturally a liquid. When cooled to freezing, it becomes ice (a solid), and when heated to boiling, it becomes steam (a gas). The simplest form of any matter is called an **element**. The element is the smallest form of matter in which the unique characteristics of the substance can still be identified. In 1869, Dmitri Ivanovich Mendeleev and Julius Lothar Meyer assembled and published all the elements known at that time into a structured table known as a periodic table of the elements. Figure 1–2 is a table of all of the natural and some of the man-made elements.

The smallest piece of an element, that still has the characteristics of the element, is called an **atom**.

Atoms have three main parts: electrons, protons, and **neutrons**. The proton and neutron combine to form the atom's nucleus. Smaller subparticles of the proton and neutron are called **quarks**. Figure 1–3 illustrates a simple form of an atom—one electron, one proton, and one neutron. Recall that the electron has a negative charge and the proton a positive charge.

Note that the neutron is neutral—it has no charge. This means that the nucleus, which is made up of the proton and neutron, is positively charged.

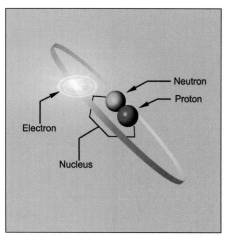

FIGURE 1–3   Hydrogen atom—the simplest atom.

You can tell the type of element by the number of protons or positive charges in the atom's nucleus. For example, silver has 47 protons in its nucleus, iron 26, and oxygen 8. The number of protons also equals the element's atomic number on the periodic table of the elements (Figure 1–2). Not all elements have a nucleus that has the same number of protons and neutrons. For example, carbon has a nucleus that contains 6 protons and 6 neutrons. Copper, however, has a nucleus that has 35 neutrons and only 29 protons (see Figure 1–4).

Some materials may come in several different forms with the same number of protons but different numbers of neutrons. These materials are called **isotopes**. One of the best-known examples of these is carbon. As stated before, carbon usually has six protons and six neutrons. This common isotope of carbon is called carbon-12 and is abbreviated as C^{12}. Another very useful isotope of carbon has six protons and eight neutrons—carbon-14 or C^{14}. Fortunately, all the various isotopes of elements behave in the same way electrically.

1.3 Atoms and Lines of Force

The electron and proton have two characteristics besides charge: size and mass (heavy or light). The electron is about three times larger than the proton but is much lighter. Even though it is only one-third the size of the electron, the proton has a mass that is over 1,836 times greater. A practical example would be comparing a 15-inch-diameter balloon to a 5-inch-diameter lead cannonball. Atoms that have the same number of electrons as they do protons are considered electrically balanced

FIGURE 1–4 Copper and carbon (C^{12}) atoms.

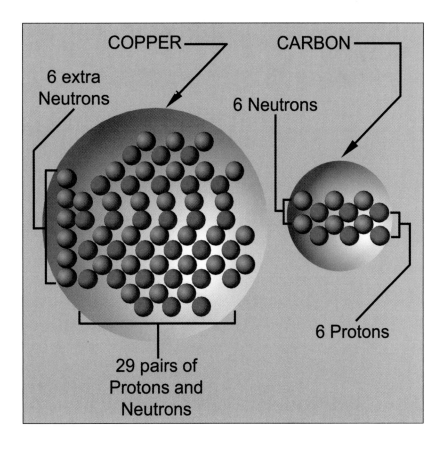

COPPER   CARBON

6 extra
Neutrons

6 Neutrons

29 pairs of
Protons and
Neutrons

6 Protons

FIGURE 1–5 Lines of force.

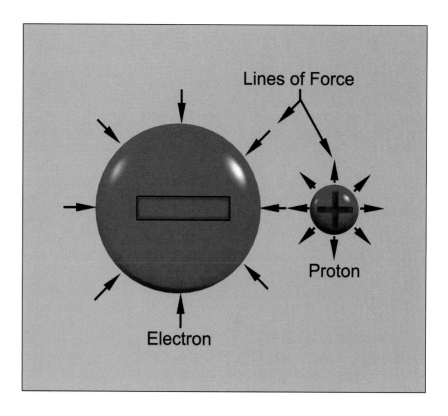

or electrically neutral in nature. This neutral condition means that opposing forces or lines of force are exactly balanced, without any effect either way. Take a look at Figure 1–5. When you look at the proton and the electron, you see that each of them has lines of force. The difference between the proton and the electron is the direction of those lines of force. The electron's lines of force "flow" inward toward the electron. This inward flow is caused by the negative charge of the electron. The proton is just the opposite. Since the proton has a positive charge, the lines of force of the proton "flow" outward in all directions.

When you need to use these electrical lines of force, work must be done to separate the electrons and protons. Changing the balance of forces produces evidence of electricity. Again, remember Benjamin Franklin's "convention." Unlike charges attract, like charges repel. This means that electrons are attracted or pulled toward protons and that protons repel or push away from other protons (see Figure 1–6).

1.4 The Forces That Bind

The Strong Nuclear Force

So what keeps all the protons and neutrons together in the center of the atom? Since the protons all repel each other, why don't they simply fly apart? The accepted theory states that a **strong nuclear force** holds the nucleus together. This force is carried by another particle called the **gluon** and is also responsible for binding the *quarks* together to form the protons and neutrons.

FIGURE 1–6   Attracting lines of force.

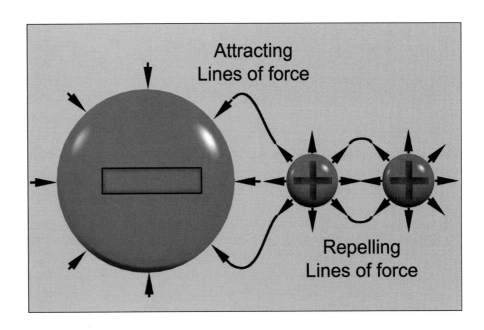

Centrifugal Force

The next logical question is, What keeps the electrons from falling into the nucleus since they are attracted by the protons? As the electrons rotate about the nucleus, there is a force that tries to cause them to fly off into space. This force is called the *centrifugal force* and exactly balances the force trying to pull them into the nucleus. The centrifugal force is also the force that keeps the string tight on a model airplane as it whirls about the operator. Figure 1–7 illustrates the concept of centrifugal force.

1.5 Electrons and Their Orbits

Although there are many possible ways in which electrons and protons might be grouped, they come together in very specific combinations that produce stable arrangements (atoms). The simplest form is

FIGURE 1–7   An example of centrifugal force.

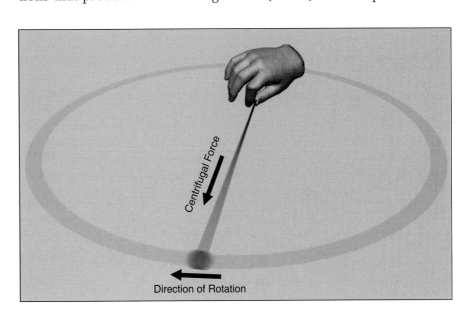

FIGURE 1–8 Electrons in their orbits.

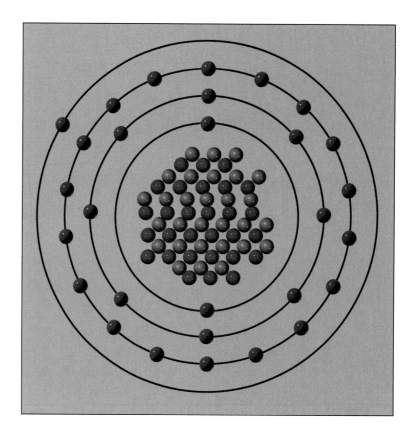

the hydrogen atom (Figure 1–3), which contains one proton and one electron, the electron spinning in "orbit" around the nucleus. There are a number of orbital rings (shells) around a nucleus that can hold electrons. Figure 1–8 shows an atom with a number of rings filled with electrons in orbit around the atom's nucleus.

The number of electrons each shell or orbital ring can hold is determined by the formula $2N^2$. N stands for the number of the shell or ring. Follow the examples given here to calculate the number of electrons that each shell or orbital ring can hold.

For the first shell or orbital ring (the number in parentheses stands for N, or the number of the shell or orbital ring),

$$2N^2 \text{ or } 2 \times (1)^2 \text{ or } 2 \times 1 = 2$$

For the second shell or orbital ring,

$$2N^2 \text{ or } 2 \times (2)^2 \text{ or } 2 \times 4 = 8$$

For the third shell or orbital ring,

$$2N^2 \text{ or } 2 \times (3)^2 \text{ or } 2 \times 9 = 18$$

For the fourth and fifth shell or orbital ring (these rings can hold a maximum of 32 electrons),

$$2N^2 \text{ or } 2 \times (4)^2 \text{ or } 2 \times 16 = 32$$

FIGURE 1–9   Valence ring (shell) and valence electrons.

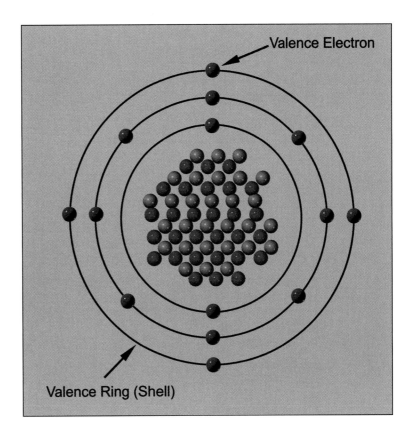

FIGURE 1–9   Valence ring (shell) and valence electrons.

1.6 Valence Shell and Electrons

Of particular interest in the study of electricity are the electrons in the outer ring, the *valence ring* of the atom. The valence ring can contain a maximum of eight electrons. These electrons are called **valence electrons** and can be easily freed in those materials that have few electrons in this outer orbit. Materials with a partially filled valence ring are chemically unstable. Because of this unstable condition, less energy is required to remove electrons from their orbits. The concept of a valence ring and electron is shown in Figure 1–9.

■ ELECTRICAL PROPERTIES OF MATERIALS

1.7 General

All materials will conduct electricity to some degree; however, some materials conduct quite readily, and some do not conduct well at all. A material that readily passes electrical current, is called a **conductor**, a material that strongly resists the flow of electricity is called an **insulator**, and a material that falls toward the middle of the conductivity scale is called a **semiconductor**. Where a material falls on this scale is determined in part by the number of electrons in their valence ring.

1.8 Conductors

Conductors are good examples of materials whose atoms have only one or two electrons in the valence ring. Examples of good conductors

FIGURE 1–10 Copper atom.

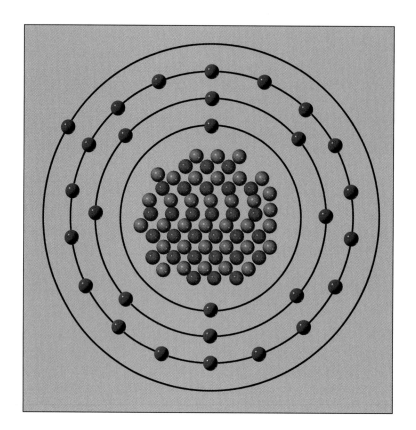

include silver, gold, platinum, copper, and aluminum. Silver, gold, platinum, and copper all have one valence electron, but silver conducts best of these metals. Aluminum has three valence electrons (three electrons in the valence shell) and is a better conductor than platinum, which has only one valence electron. Remember that the easier it is for the electron to be pulled away from its valence orbit, the better the atom "conducts." Figure 1–10 is a picture of a copper atom, one of the most common conducting metals.

A material is said to "conduct electricity" when one electron of an atom is forced from its orbit by another atom's electron. When an atom has only one valence electron, it can be easily bumped or forced from orbit by another electron. When an electron impacts (strikes) another electron, the electron being hit takes energy from the striking electron. This extra energy forces the "hit" electron to leave orbit and jump to a neighboring electron's orbit, and the process is repeated. The electron that impacted the first valence electron now takes the leaving electron's place as the new valence electron. Figures 1–11a and b detail this "electron flow."

An important fact to note here is that not all the energy is transferred when the electrons collide. Some of the energy is lost in the form of resistance. This resistance comes from the original valence electron's not wanting to leave its orbit. The energy expended in moving electrons is released as heat (see Figure 1–11a). That is why wires that conduct electricity get warm. Too much electron flow (electrical current) can overheat the wires and their insulation and may ignite.

When one electron strikes a **valence shell** with two valence electrons, the energy of the incoming electron is divided between the two valence electrons. This is illustrated in Figure 1–11b.

FIGURE 1–11   Electron flow:
(a) one electron in valence ring;
(b) two electrons in valence ring.

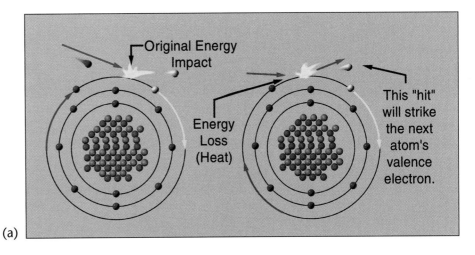

(a)

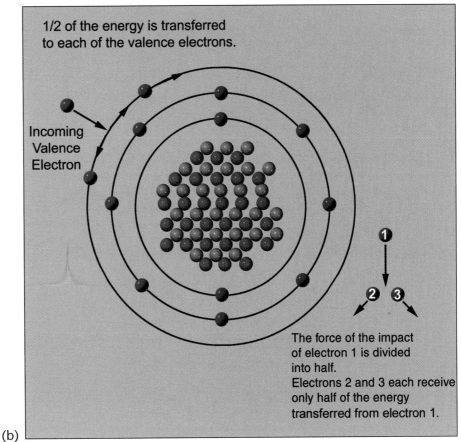

(b)

In electrical circuits, the electrons travel through the electrical circuit. The force of attraction or repulsion of the electrons in a circuit is fundamentally a measure of the quantity of the electron's charge. This ability or "potential" to attract or repel other electrons refers to the possibility of doing work. Any charge has the potential to do the work of repelling or attracting other charges. When you talk about two unlike charges (a positive and a negative), you have a potential difference.

The flow of these electrons is always from the pole of an electrical energy source that has an excess of electrons to the pole that has a deficiency of electrons. This is in compliance with the atom attempting

FIGURE 1–12 Electrons flow from an excess to a deficiency.

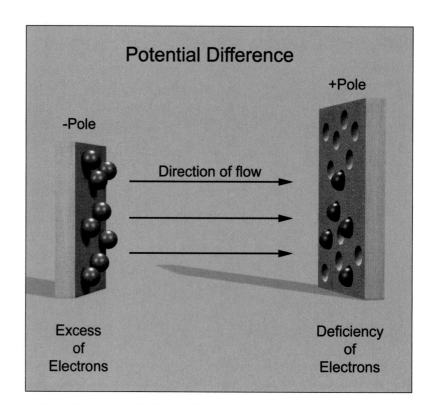

to reestablish electrical balance. The direction of current flow, negative to positive, is defined as the polarity of the circuit. This is an important concept in the study and understanding of many electrical principles. Figure 1–12 shows how this electron flow works.

1.9 Insulators

Materials with atoms in which the electrons have a very high resistance to leaving their own valence rings are known as insulators. These materials cannot conduct very well because their electrons do not readily move from atom to atom, as seen in Figure 1–11a. The valence shell or ring that is almost full (seven or eight valence electrons) is very stable; these electrons require a tremendous amount of energy to break free of their orbits. Any incoming electron gives up its energy to all the electrons in the valence ring. These electrons do not break free but do store the incoming energy, as in Figure 1–13.

Examples of good insulator materials include rubber, glass, plastic, paper, wood, air, and mica. Insulators are useful when it is necessary to prevent current flow. In addition, they can also be used to store an electrical charge.

1.10 Semiconductors

Semiconductors conduct more poorly than metal conductors but better than insulators. Semiconductors have four electrons in the outermost ring (the valence shell). This means they neither gain nor lose electrons but share them with other similar atoms. The reason for this is that four is exactly one-half of the stable condition of eight electrons in the valence ring. Figure 1–14 is a diagram of a semiconductor.

FIGURE 1–13 Incoming
electrons in an insulator.

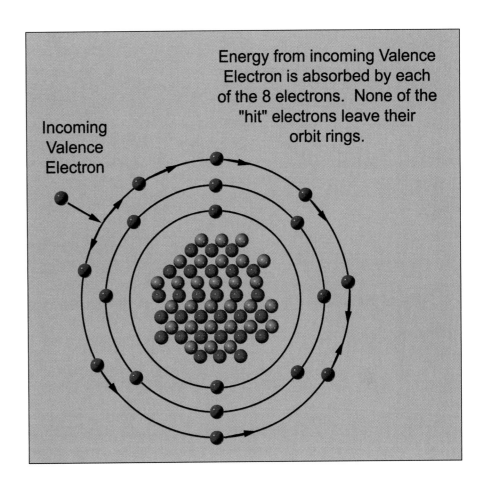

FIGURE 1–14 Semiconductor
valence ring.

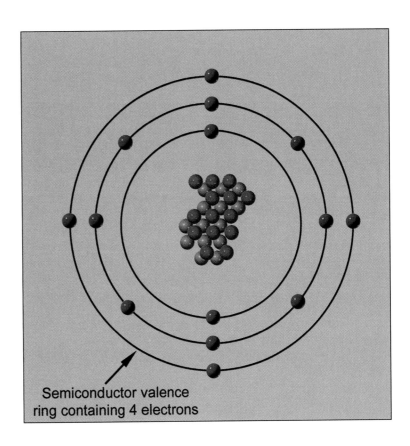

Since 1947, when the transistor (a semiconductor device) was invented, semiconductors have been used in almost all phases of our life. Solid-state or semiconductor devices are used in everything from radios, televisions, toasters, and pagers to motor controls and garage door openers. Some of the most common semiconductor materials are carbon, silicon, and germanium. Silicon is most often used in diodes, transistors, and integrated circuits. Before a "pure" semiconductor material such as silicon can be used in a device such as a transistor, it must be "doped" (mixed) with an impurity.

The addition of heat causes changes in the conductivity of all materials. When an insulator or a semiconductor is heated, its resistance tends to go down; consequently, it becomes a better conductor.

When a conductor is heated, its resistance goes up. This information will be useful in understanding wire sizes and current flow.

■ MOLECULES, COMPOUNDS, AND IONS

1.11 Molecules and Compounds

A group of two or more atoms forms a molecule. For example, two hydrogen (H) atoms make a hydrogen molecule (H_2). When this hydrogen molecule combines chemically with an oxygen (O) atom, you have a water ($H_2 + O$ or H_2O) molecule. Compounds are made up of two or more elements, (such as oxygen, hydrogen, copper, or silicon). A molecule is the smallest unit of a compound with the same chemical characteristics. You can have molecules for elements as in the example of two hydrogen atoms combining (H_2) or for compounds such as H_2O. See the examples in Figure 1–15.

FIGURE 1–15 Hydrogen and water molecule.

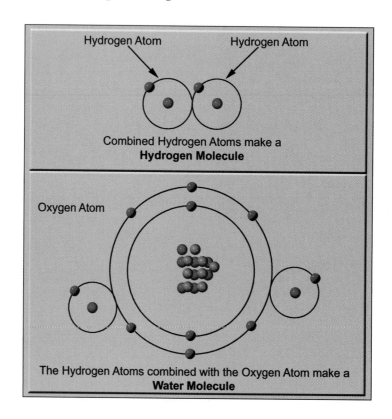

1.12 Ions and Their Electrical Use

Ions are atoms that have either gained or lost electrons. An atom that has lost electrons is a positive ion. An atom that has gained electrons is a negative ion. The process of an atom losing or gaining electrons is called *ionization.* Negative ions (atoms) have more electrons than protons. Positive ions (atoms) have fewer electrons than protons.

Remember that electrical current (electron) flow is caused by a potential difference between two poles. One pole has an excess of electrons (the negative pole), and the other has a deficiency of electrons (the positive pole—see Figure 1–12). Normally you think of this current flow as moving through a wire. However, depending on the atoms, element structure, and molecules, this electrical current could be through a gas or a liquid. In gases or liquids, the electrical flow is caused by individual ions rather than individual electrons.

An example of two different elements' atoms (chlorine and magnesium) will show the effects of ionization. The element chlorine is not considered a metal because it has seven valence electrons (electrons in the outermost valence shell). Magnesium is considered a metal because it has two valence electrons in its outermost valence shell (see Figure 1–16.)

When magnesium is mixed with chlorine gas and heated, the two elements (one atom of magnesium and two atoms of chlorine) combine to form a molecule of metallic salt called magnesium chloride. Figure 1–17 shows a diagram of the combination results.

These atoms are now called ions. The magnesium atom is a positive ion because it has lost two electrons, and the chlorine atoms are negative ions because they have each gained an electron from the magnesium atom. Another molecule that combines a metal element atom (sodium) with chlorine in this same way is sodium chloride, or salt.

1.13 Electrolytes

When a salt such as sodium chloride (table salt) is dissolved in water, the ionized particles will allow the passage of electricity. Figure 1–18 shows how a glass of pure water (a nonconductor) can be transformed

FIGURE 1–16   Magnesium and chlorine atoms.

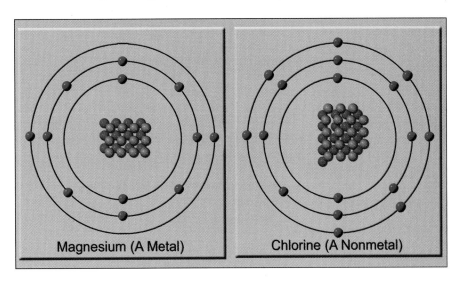

FIGURE 1–17 Molecule of the metallic salt magnesium chloride.

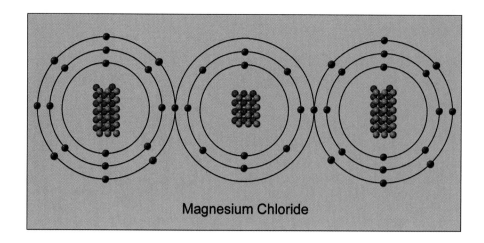

Magnesium Chloride

FIGURE 1–18 Lowering the resistance of water by creating an electrolyte solution.

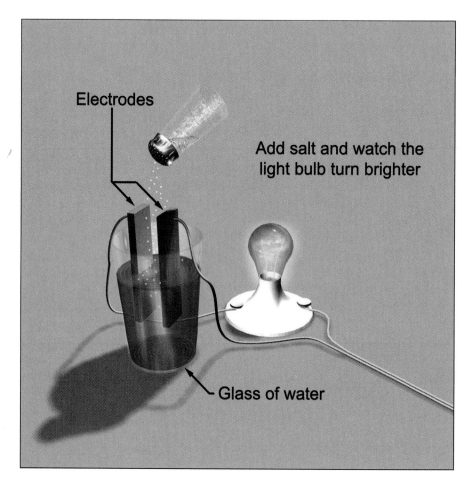

Electrodes

Add salt and watch the light bulb turn brighter

Glass of water

into a conductor by dissolving salt in the water. Note that the electrodes shown in Figure 1–18 are made of the same material—copper, for example. Acids and alkalies can also be used to cause electrical conduction in liquids. These types of solutions are called **electrolytes**. Refer back to Figure 1–18.

When two unlike metals are placed in an electrolyte solution, an electrochemistry process takes place that causes the transfer of electrons—through the electrolyte—from one of the metals to the other. The metal that accumulates electrons is the negative pole, and the metal that gives up electrons is the positive pole. In this manner, an electrical potential is developed between the two poles. This action will be discussed again when you learn about batteries.

1.14 Other Useful Ions

Other types of materials also have useful electrical functions. One of the more useful is copper sulfate, which is used in the copper electroplating process. Its symbol is $CuSO_4$. When mixed with water, two ions are created: a negative sulfate ion (SO_4), which has two extra electrons from the copper atom, and Cu, a cupric ion, which is positive because it has lost two electrons to the sulfate ion.

Sulfuric acid is also a compound made of two ions. Its main use is the electrolyte liquid used in lead-acid car batteries. Sulfuric acid's symbol is H_2SO_4. This indicates that sulfuric acid is made up of two hydrogen atoms, one sulfur atom, and four oxygen atoms. When H_2SO_4 is combined with water, its molecules separate into three ions—SO_4, H, and H. The two hydrogen atoms (H) are positive because they have lost their valence electron to the SO_4 ion. The SO_4 ion is negative because it now has two extra electrons.

■ SOURCES OF ELECTRICAL ENERGY

1.15 Sources and Effects

We have been describing the makeup of atoms and how electrons can flow or be forced from one atom to another. This causes different effects and combinations to happen. Many of these effects have useful electrical purposes. What causes the electrons to leave their valence shells and "hit" or take up orbit in another atom's valence ring? The answer is in what causes the initial electron to be forced from its shell.

There are six methods that are known to force electrons from the valence ring of an atom to become a potential electrical current participant:

1. Friction
2. Chemical
3. Heat
4. Pressure
5. Light
6. Magnetism

With the exception of friction, electrical energy can produce the same effects as those used to produce it. That is, if the proper load is connected across a source, any one (or more) of the following effects will occur:

1. Chemical action
2. Heat
3. Pressure
4. Light
5. Magnetic fields

1.16 Friction

Friction was the first known method of generating electricity. Its discovery is usually credited to the Greeks. This method consists of rubbing two pieces of material together, such as a rubber rod and wool or silk and a glass rod. The friction created by rubbing the objects together causes electrons to be transferred from one of the materials to the other. The change that is created is called **triboelectricity.** Friction is a source of **static electricity.** Static means that the electricity is merely stored and is at rest. There is no movement of electrons after the transfer is complete. The static charge is potential electrical energy that will cause the excessive electrons on one body to flow to a body with a deficiency of electrons until they are electrically balanced. In nature, the tendency is always to establish an electrical balance. The body with excessive electrons is said to be negative, and the body with a deficiency of electrons is said to be positive.

1.17 Chemical and Chemical Action—Primarily Batteries

The chemical source of electricity is best represented in our everyday lives by the **battery.** These devices are used to start our cars, power portable systems, and operate emergency lighting and other systems when our primary source of electrical power fails. Batteries can be large (a car or truck battery), or they can be small enough to fit in the back of a wristwatch. The battery is a primary source of the direct current (DC) charge.

Battery History

The battery (electricity produced by chemical action) was first discovered by Luigi Galvani in 1791. He was conducting anatomy experiments using frog legs that were preserved in a salt solution. He discovered that when the legs were suspended by copper wires and touched by an iron scalpel, they would "twitch." He knew this twitch was caused by electricity, but he mistakenly thought the electricity was caused by the contracting muscles of the frog.

Alessandro Volta repeated Galvani's experiment in 1800. He correctly identified the cause of the electricity as a combination of the copper wire, iron scalpel, and salt solution. Volta continued his experiments until he developed the first battery. It was made from zinc and silver disks held apart by pieces of cardboard that had been saturated with salt water. This single structure with one silver disk, one zinc disk, and one piece of saturated cardboard is called a **cell.** When many of these cells are connected in series or parallel, the resulting structure is called a battery. Volta called his "battery" a voltaic pile because it was a series of stacks that were connected together. Figure 1–19 shows the symbol for a cell and for a battery.

Battery Construction

An easy-to-construct example of a cell is shown in Figure 1–20. Connect a piece of aluminum wire to one side of a voltmeter and a copper wire to the other side. Insert the other ends of the aluminum and cop-

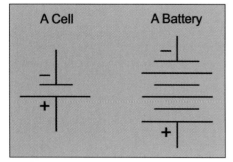

FIGURE 1–19 Cell and battery symbols.

FIGURE 1–20   A simple "potato cell."

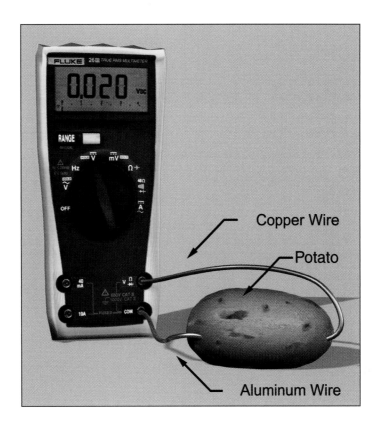

per wire into each end of a raw potato. The acid in the potato acts as the electrolyte, and current will flow between the two unlike metals (aluminum and copper). You should be able to see a small voltage register on the voltmeter.

An even stronger electrical cell can be constructed using a copper wire, a galvanized roofing nail, and a grapefruit.

Battery Metals

Individual cell voltage depends on what metals are used for each plate. A list of special metals called the "Electromotive Series of Metals" is shown in Table 1–1. This list is not complete, but provides good examples listed in the order in which they can receive or give up electrons. The first metal, lithium, receives electrons more easily than the rest. Notice the distance between aluminum and silver. Since aluminum will receive electrons more easily than silver, when paired together aluminum will be the negative electrode and silver the positive electrode.

As a practical matter, not all these materials will make good metal for battery cells. Many will corrode, and others will cause chemical reactions in the electrolyte. Either of these problems will cause resistance to electron flow and reduce the cell's ability to produce electricity.

Battery Categories

Batteries are usually divided into two categories: primary cells and secondary cells. **Primary cells**, like most batteries used in portable radios and flashlights, cannot be recharged. Once they have depleted

Table 1–1   Electromotive Series of Metals

Lithium
Potassium
Sodium
Magnesium
Aluminum
Manganese
Zinc
Iron
Cobalt
Nickel
Tin
Lead
Copper
Mercury
Silver
Platinum
Gold

Table 1–2 Primary and Secondary Cells

Cell	Positive Plate	Negative Plate	Electrolyte	Volts per Cell
Secondary Cells				
Lead-acid	Lead dioxide	Lead	Diluted sulfuric acid	2.2
Nickel-iron	Nickel dioxide	Iron	Potassium hydroxide	1.4
Nickel-cadmium	Nickel hydroxide	Cadmium	Potassium hydroxide	1.2
Silver-cadmium	Silver oxide	Cadmium	Potassium hydroxide	1.1
Primary Cells				
Carbon-zinc	Carbon manganese dioxide	Zinc	Ammonium chloride	1.5
Alkaline	Manganese dioxide	Zinc	Potassium hydroxide	1.5
Mercury	Mercuric oxide	Zinc	Potassium hydroxide	1.35
Zinc-air	Oxygen	Zinc	Potassium hydroxide	1.4

their chemical action, they must be thrown away. **Secondary cells**, like those used in automobiles, can be recharged several times. Table 1–2 shows a chart of primary and secondary cells.

An example of a secondary cell is the automobile battery. When such a battery is recharged, the chemical action that allows electric current to be drawn from the battery is reversed. Passing electric current through a water and sulfuric acid solution will separate the water molecules into two hydrogen ions and one oxygen ion. The positive hydrogen ions will attract negative sulfate ions and produce sulfuric acid. The negative oxygen ions are attracted by the positive electrode. Therefore, as the battery charges, more and more of the electrolyte is changed from water to sulfuric acid, increasing the battery's charge.

Besides batteries, other chemical actions, such as electrolysis, are caused by an electrical current passing through certain materials. Electroplating gold jewelry is a good example of this chemical action.

1.18 Heat

Heat is generated any time current flows through a wire. This happens because energy is used to move electrons. Usually, the effect of heating a wire is undesirable. For this reason, metals such as copper and aluminum are normally used as conductors. For certain applications, heat can be a desirable outcome. Such products as toasters, irons, clothes dryers, and space heaters are designed to produce heat for practical purposes.

A transfer of electrons can also take place when two dissimilar metals are joined together at a junction and then heated. This is known as the **thermoelectricity** process. For some active metals, the ambient room temperature, at normal comfort level, is sufficient to cause the transfer of electrons from one metal to the other. Increases in the temperature cause a greater transfer of electrons. This type of device is called a **thermocouple**.

1.19 Pressure

Certain crystalline substances, when placed under pressure, generate minute potentials (potential differences). The force of the pressure causes the electrons to be driven out of orbit in the direction of the force. The electrons leave one side of the material and collect on the other side. Depending on how the crystal is cut, some respond best to bending pressures, while others react best to a twisting action. Electricity derived from pressure is known as the **piezoelectric effect**.

It is possible to reverse the process and produce pressure with electrical current. The pressure is produced by the physical displacement of the ions in the crystal.

1.20 Light

Light Producing Electricity

Light has small particles of energy called *photons.* When photons strike certain types of photosensitive material, they release energy into the material. There are three types of photoelectric effects of interest in the study of electricity:

1. **Photoemission.** When photons strike the surface of the material, electrons are released in a vacuum tube. A positive plate placed in the tube will cause the electrons to flow to it. More electrons will flow with greater light energy.
2. **Photovoltaic.** When photons strike one of two photosensitive plates that are joined together, electrons move from the plate that is bombarded by the light energy to the adjacent plate. An electrical potential then exists between the two plates. This action is similar to that of a battery, except light energy is used rather than chemical energy. This component is called a *photocell.*
3. **Photoconduction.** When photons bombard some types of materials that are normally poor conductors, electrons are freed from the valence shell to participate in electrical current flow. This in turn lowers the resistance of the material.

Electricity Producing Light

When enough electrical current is passed through a poor conductor, not only is heat generated, but some materials will also begin to glow red or even white-hot. Everyone has seen the red glow of a burner on an electrical stove. The common electrical incandescent lamp is designed to give off white light when it is heated to incandescence. The heat given off with this process is usually considered an energy loss or "heat loss."

There are four other methods of producing light with electricity that do not result in as much heat loss as in the incandescent lamp:

1. **Electroluminescence.** Some materials, such as neon gases, argon, and mercury vapor, give off light when current is passed through them. Because the usable amount of light is small, electroluminescence methods are used mostly for displays, such as neon signs.

2. **Phosphorescence.** Light occurs when an electron beam strikes a surface covered with phosphors. This is how light is provided on the screen of a cathode ray tube used in television sets and oscilloscopes.

3. **Fluorescence.** Fluorescent lamps make use of both electroluminescence and phosphorescence to provide a higher level of light output than that of an incandescent lamp while using an equal amount of electrical power. These lamps have a gas, such as mercury vapor, that becomes an ionized carrier of electrical current. This ionization process causes ultraviolet radiation that strikes the phosphorescent coating on the inside of the fluorescent tube, causing "white light."

4. **pn junction luminescence.** This type of luminescence occurs when direct current is applied to a pn junction. The pn junction must be specially doped with other materials, such as the semiconductors discussed earlier in this text. When electricity passed through the doped pn junction, electrical energy is absorbed by the junction. This electrical energy is then released in the form of light and heat energy. These special pn junctions are called light-emitting diodes (LEDs). LEDs can be manufactured that will produce any color of light, from infrared to near ultraviolet. Individual LEDs do not generate a large amount of light but are used extensively as indicator lights, in numeric displays, and in specialty equipment, such as signs and long-life emergency exit light lamps.

1.21 Magnetic Fields and Magnetism

Magnetism from Electricity

Magnetism is the primary source of electrical power. Any time current flows through a conductor, not only is heat generated, but also a magnetic field. A large current causes a stronger field than a small current.

The presence of the magnetic field can be detected using a small magnet, such as the one in a compass (see Figure 1–21). When current flows through a conductor that is close to the compass, the needle will be deflected from pointing north.

Electricity from Magnetism

When a conductor moves through a magnetic field, electrons will move from one end of the conductor to the opposite end, creating a potential difference between the two ends. Only relative motion is necessary between the magnetic field and the conductor. The conductor can be stationary and the magnetic field moving or vice versa. If the magnetic field and conductor are moved together at the same rate, no electrical energy will be generated because there is no relative motion between the two.

This process of producing electrical energy is called magnetoelectricity and makes possible electrical magnets, motors, generators, and transformers. In fact, most commercial electricity is generated using this method.

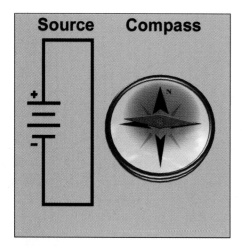

FIGURE 1–21   Magnetic field created by current flow.

■ SUMMARY

The principles of electricity are based on a modern concept called the *electron theory of electricity.* This theory attributes all electrical phenomena to small atomic particles called *electrons.* These electrons, along with protons and neutrons, make up atoms—the smallest piece of an element. Atoms can be thought of as small solar systems with neutrons and protons at the center (nucleus) and electrons circling in concentric shells around the nucleus.

Electrons have negative charges, while protons have positive charges. All electrical activity and effects can be attributed to the concentration, motion, and interaction of electrons and protons. Some materials, called *conductors,* allow electrons in their outermost shell (valence shell) to move freely, while other materials, called *insulators,* have the electrons locked in place. By proper use of such materials, electrical energy can be transferred from its sources—batteries and generators, for example—to its end uses—lights, motors, heaters, and the like.

The effects of electrons in motion are described in more detail in later chapters. In this chapter, you learned that electrons in motion can create heat, light, and magnetism. Such effects can be used in the end-use devices to perform useful work.

■ REVIEW QUESTIONS

1. What are the three particles that make up atoms?
2. Which particle has a negative charge? Which has a positive charge?
3. What is the valence shell of an atom?
4. Which shell of electrons is the primary determinant of the electrical behavior of a material?
5. How many electrons are found in the valence shell of a sodium (Na) atom?
6. When salt (NaCl) is dissolved in water, which substance is the positive ion, and which is the negative ion?
7. Electron current flow is always from a ____ concentration of electrons to a ____ concentration of electrons.
8. How does a battery store electrical energy?
9. Generators use what other type of naturally occurring force to create electricity?
10. How can energy be transferred using electrons?

Mathematical Concepts for Solving Electrical Problems

■ OUTLINE

■ OVERVIEW

During your career as a practicing electrician, you will be called on to perform mathematical calculations. Most of these calculations will be fundamental operations, such as addition, subtraction, multiplication, or division. Occasionally, however, electricians are required to perform algebraic manipulations of Ohm's law, wire resistance formulas, and other such formulas.

In this chapter, you will begin to apply the basic principles of algebra to the equations and formulas you will be studying and using throughout your career. The focus of this chapter is on solving equations in which there is only *one unknown quantity*. The ability to solve such equations may be crucial to your individual success and understanding of the fundamental relationships between current, voltage, resistance, power, and many other such variables in a circuit. As your competency expands, you will become ready, willing, and able to tackle ever more complex calculations.

■ OBJECTIVES

After completing this chapter, you should be able to:

1. State from memory the prefix multipliers and their meanings.
2. Write simple equations needed to solve for unknown values.
3. Solve simple algebraic equations for one unknown.
4. Rearrange equations to isolate an unknown variable.

■ GLOSSARY

Equation A statement asserting the equality of two expressions, usually written as a linear array of symbols that are separated into left and right sides and joined by an equals sign.[1]

Equivalent equation The equation that results when the unknown value is isolated.

Formula A statement, especially an equation, of a fact, rule, principle, or other logical relation.[2]

Inverse Something that is opposite, as in sequence or character; the reverse.[3]

Isolating A key step in solving an equation for an unknown variable. An unknown is isolated by moving it to one side of the equation. All other values are moved to the other side. Of course, the rules of algebra must be followed during this process.

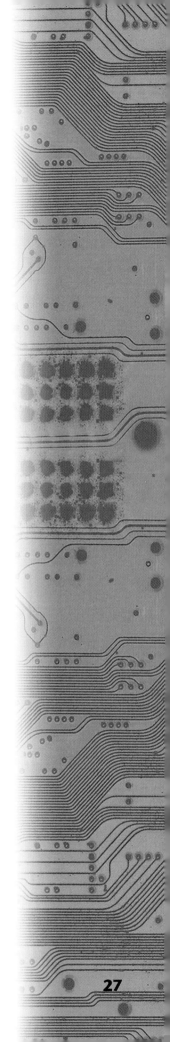

■ INTRODUCTION

In this chapter you will begin to apply the basic principles of algebra to the equations you will be working with throughout your career. The focus of the chapter will be on solving equations you work with in your career in the electrical business.

■ PREFIX MULTIPLIERS

Electrical and electronic component and circuit values can be extremely small or large. Prefix multipliers are used to simplify the printing and use of small or large figures. For example .001 is expressed using the prefix *milli;* thus, .001 amperes may be expressed as one *milliampere.*

The method of applying such prefixes is straightforward. Divide the original value by the number value of the prefix and add the prefix. For example,

$$2,000 \; amperes = \frac{2,000}{1,000} \; amperes = 2 \; kiloamperes$$

Note that prefix multipliers are sometimes abbreviated even further. *2 kiloamperes* is usually expressed as *2 kA.* Table 2–1 shows the values, names, and powers of ten for the most commonly used prefixes.

■ WRITING EQUATIONS

Many electrical problems are solved by solving an algebraic **equation**. Of course, before an equation can be solved, it must first be written.

Table 2–1   Common Prefixes and Their Values

Number Value	Metric Prefix	Power of Ten
1 000 000 000 000	Tera	10^{12}
1 000 000 000	Giga	10^{9}
1 000 000	Mega	10^{6}
1 000	Kilo	10^{3}
100	Hecto	10^{2}
10	Deka	10^{1}
1	Units or one	10^{0}
0.1	Deci	10^{-1}
0.01	Centi	10^{-2}
0.001	Milli	10^{-3}
0.000 001	Micro	10^{-6}
0.000 000 001	Nano	10^{-9}
0.000 000 000 001	Pico	10^{-12}

Sometimes a number of facts concerning a given problem are known. Subsequently, it is up to the electrician to develop an equation that will produce the correct solution.

Writing equations is not as difficult as you might think. What it really amounts to is simply taking phrases or sentences that are written in English and expressing them in appropriate algebraic terms. The procedure is to substitute letters or symbols for the key words and arithmetic operations that are contained within the sentence. The following examples illustrate this procedure:

EXAMPLE 1

Sentence: Some number increased by 5 is equal to 19. ($N =$ some number)

Equation: $N + 5 = 19$

EXAMPLE 2

Sentence: 7 is 5 more than some number.

Equation: $7 = 5 + N$ or $7 = N + 5$

EXAMPLE 3

Sentence: Some number decreased by 8 is equal to 4.

Equation: $N - 8 = 4$

EXAMPLE 4

Sentence: 9 is 6 less than some number.

Equation: $9 = -6 + N$ or $9 = N - 6$

EXAMPLE 5

Sentence: Some number times 7 is equal to 98.

Equation: $7N = 98$

EXAMPLE 6

Sentence: The area of a parallelogram is equal to the length of the base multiplied by the length of the height.

Equation: $A = b \times h$ or $A = bh$

EXAMPLE 7

Sentence: Some number divided by 6 is equal to 7.

Equation: $N \div 6 = 7$ or $\dfrac{N}{6} = 7$

EXAMPLE 8

Sentence: The current (I) is equal to the voltage (E) divided by the resistance (R).

Equation: $I = \dfrac{E}{R}$

EXAMPLE 9

Sentence: Some number squared is equal to 144.

Equation: $N^2 = 144$

EXAMPLE 10

Sentence: The area of a square is equal to the length of any side squared.

Equation: $A = S^2$

EXAMPLE 11

Sentence: The square root of some number is equal to 20.

Equation: $\sqrt{N} = 20$

EXAMPLE 12

Sentence: Three times the sum of some number plus 7 is equal to 51.

Equation: $3(N + 7) = 51$

EXAMPLE 13

Sentence: The power (P) in a circuit is equal to the current (I) squared times the resistance (R).

Equation: $P = I^2 R$

EXAMPLE 14

> **Sentence:** Some number squared is equal to the sum of two other numbers squared. (You should understand here that "some number" may be represented by any letter or symbol. The letters or symbols are simply placeholders for the unknown value for which you are searching.)
>
> **Equation:** $X^2 = Y^2 + Z^2$

■ SOLVING EQUATIONS

2.1 Fundamentals

Once the equation has been written, the next step is to solve for the unknown value that it expresses. In the simple example of $x = 2 \times 3$, the equation is solved readily by completing the multiplication on the right side, producing $x = 6$.

Although the letter x is sometimes used for the unknown value, common practice frequently employs a specific letter that reminds us of the nature of the value being sought. Examples of this practice include using T for time, F for frequency, P for power, and R for resistance.

Assuming that an equation has been written correctly, the expression on the left side of the equation will be *equal* to whatever is on the right side; thus the name *equation*. In order to solve the equation, it must be rearranged so that the unknown variable is on one side (usually the left) and all other terms are on the other side. This process is called **isolating** the unknown, and all the rules of algebra must be respected. The equation that results from the isolating procedure is called an **equivalent equation**.

2.2 The Equation as a Balance Beam

When preparing to write an *equivalent equation,* it helps to think of an equation as a balance beam. The fulcrum of the balance beam is located at the equals sign (=). The left side and the right side of the beam must be kept in balance by being certain that all procedures are performed identically to the left and right sides (see Figure 2–1).

In the case of a balance beam, if 5 kilograms are removed from the left side, 5 kilograms must also be removed from the right side to keep the beam in balance.

CAUTION: Whatever you do to one side of an equation you must also do to the other side.

Table 2–2 shows examples of the proper way to perform standard mathematical functions on an equation.

2.3 Inverse Operations

To isolate the unknown and thus solve the equation, you must perform **inverse** operations on the equation. The inverse is the opposite or

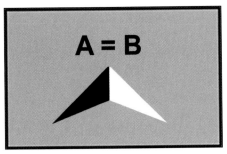

FIGURE 2–1 Equation expressed as a balance beam.

Table 2–2   Math Operations on an Equation

Action	Example
If a value is added to one side of an equation, the same value must be added to the other side.	If $A = B$, then $A + 2 = B + 2$
If a value is subtracted from one side of an equation, the same value must be subtracted from the other side.	If $A = B$, then $A - 7 = B - 7$
If one side of an equation is multiplied by a value, the other side must be multiplied by the same value.	If $A = B$, then $5A = 5B$
If one side of an equation is divided by a value, then the other side must be divided by the same value.	If $A = B$, then $A/10 = B/10$
If one side of an equation is raised to a power, the other side must be raised to the same power.	If $A = B$, then $A^2 = B^2$
If the root of one side of an equation is taken, the same root must be taken of the other side.	If $A = B$ then $\sqrt{A} = \sqrt{B}$

Table 2–3   Table of Inverse Operations

Addition	Subtraction
Multiplication	Division
Exponentiation (raising to a power)	Taking the root

complementary action. Table 2–3 shows the inverse relationships among the various types of mathematical operations.

Consider the simple equation $x + 4 = a$. To isolate the unknown variable x, you must perform the inverse operation of addition (subtraction) on both sides of the equation $x + 4 - 4 = a - 4$. This equation simplifies to $x = a - 4$. Notice that the variable x is now isolated on the left side. The rule is then: To solve an equation, isolate the variable by performing the required inverse operations to each side of the equation.

■ EXAMPLES OF EQUATION SOLVING

2.4 Fundamental Operations

EXAMPLE 1

$$X + 8 = 15$$

Solution:
Since 8 is added to X, you must subtract 8 from the right-hand side of the equation in order to isolate the variable (X). If 8 is subtracted from one side of the equation, it must be subtracted from the other side as well:

$$X + 8 - 8 = 15 - 8$$

The quantity $(+8 - 8)$ is equal to zero. This leaves X (our variable) isolated on the left-hand side of the equation. The right-hand side of the equation becomes $(15 - 8)$ or 7.

$X = 7$

Check:

$X + 8 = 15$

Determination:

$X = 7$

Therefore:

$7 + 8 = 15$
$15 = 15$

If the determined value for X is correct, the equation will balance as shown.

EXAMPLE 2

$A - 10 = 14$

Solution:
Since 10 is subtracted from the variable, the inverse operation of adding 10 must be performed on each side of the equation:

$A - 10 + 10 = 14 + 10$

Simplify:

$A = 24$

Check:

$A - 10 = 14$
$24 - 10 = 14$ (substitute for A)
$14 = 14$ (balance)

EXAMPLE 3

$27 = X + 4$

Solution:
Subtract 4 from each side of the equation:

$27 - 4 = X + 4 - 4$

Simplify:

$23 = X$ or $X = 23$

Check:

$$27 = X + 4$$
$$27 = 23 + 4$$
$$27 = 27$$

EXAMPLE 4

$$7 + X = 5$$

Solution:
Subtract 7 from each side of the equation:

Remember: ($7 + X$ is the same as $X + 7$)

$$7 + X - 7 = 5 - 7$$
$$X = -2 \text{ (answers can be negative)}$$

Check:

$$7 + X = 5$$
$$7 + (-2) = 5$$
$$5 = 5$$

EXAMPLE 5

$$23 - X = 8$$

Solution:
Subtract 23 from each side of the equation:

$$23 - X - 23 = 8 - 23$$
$$-X = -15$$

Although this is a correct solution, it is not in proper form. The variable should always appear as a positive. To change $-X$ to $+X$, proceed as follows. $-X = -15$ is the same as $(-1)(X) = (-1)(15)$. Since X is now multiplied by -1, do the inverse operation of dividing each side of the equation by -1:

$$\frac{(-1)(X)}{-1} = \frac{(-1)(15)}{-1}$$

The -1s cancel each other out, leaving $X = 15$.

The negative variable can also be approached in another way. Remember that any time a negative number multiplies a negative number, the result is a positive number. This principle often utilizes -1, so that the absolute value of the number is not changed with the sign:

$$(-) \times (-) = (+)$$
$$(-1) \times (-1) = (+1)$$
$$(-X) = (-15)$$
$$(-1)(-X) = (-1)(-15)$$
$$X = 15$$

Check:

$$23 - X = 8$$
$$23 - 15 = 8$$
$$8 = 8$$

EXAMPLE 6

$$7X = 35$$

Solution:
X is multiplied by 7, so divide each side of the equation by 7:

$$\frac{7X}{7} = \frac{35}{7}$$
$$X = 5$$

Check:

$$7X = 35$$
$$(7)(5) = 35$$
$$35 = 35$$

EXAMPLE 7

$$\frac{X}{4} = 20$$

Solution:
X is divided by 4, so multiply each side of the equation by 4:

$$\frac{4 \times X}{4} = 20 \times 4$$

The 4s on the left side of the equation cancel each other:

$$\frac{\cancel{4} \times X}{\cancel{4}} = 20 \times 4$$
$$X = 20 \times 4 \text{ (right side of the equation)}$$
$$X = 80$$

Check:

$$\frac{X}{4} = 20$$

$$\frac{80}{4} = 20$$

$$20 = 20$$

EXAMPLE 8

$$-5Y = 20$$

Solution:
Divide each side of the equation by -5:

$$\frac{-5Y}{-5} = \frac{20}{-5}$$

$$Y = -4$$

Check:

$$-5Y = 20$$

$$-5(-4) = 20$$

$$20 = 20$$

EXAMPLE 9

$$\frac{-C}{15} = 4$$

Solution:

$$\frac{C}{-15} = 4$$

Multiply each side by -15:

$$\frac{(-15)C}{-15} = 4(-15)$$

$$C = -60$$

Check:

$$\frac{-C}{15} = 4$$

$$\frac{-60}{-15} = 4$$

$$4 = 4$$

EXAMPLE 10

$$X^2 = 225$$

Solution:

X is squared; therefore, do the inverse operation of taking the square root of each side of the equation:

$$\sqrt{X^2} = \sqrt{225}$$
$$X = 15$$

Check:

$$X^2 = 225$$
$$15^2 = 225$$
$$225 = 225$$

EXAMPLE 11

$$\sqrt{X} = 8$$

Solution:

Square each side of the equation to isolate X:

$$(\sqrt{X})^2 = 8^2$$
$$X = 64$$

Check:

$$\sqrt{X} = 8$$
$$\sqrt{64} = 8$$
$$8 = 8$$

2.5 Multiple Operations

As equations become more complex, it is often necessary to perform more than one operation on each side of the equation, in order to isolate the variable. There are no definite rules as to which operation should come first. However, doing certain operations first may make the solution easier to find and easier to understand. The following examples will provide you with several tips on which operation should come first. This information can be extremely useful.

EXAMPLE 1

$$7X + 3 = 59$$

Solution:

First, subtract 3 from each side of the equation. In this problem, the number 3 represents what is referred to as a "loose constant"—it is easily moved to the other side of the equation. Here, that movement is achieved through subtraction. At this point in the solution, it is much easier to subtract 3 from each side of the equation than it is to divide each side by 7:

$$7X + 3 - 3 = 59 - 3$$

Simplify:

$$7X = 56$$

Next, divide each side by 7:

$$\frac{7X}{7} = \frac{56}{7}$$

Simplify:

$$X = 8$$

Check:

$$7X + 3 = 59$$
$$7 \times 8 + 3 = 59$$
$$56 + 3 = 59$$
$$59 = 59$$

EXAMPLE 2

$$\frac{3X}{4} = 6$$

Solution:

First, multiply each side of the equation by 4. Fractions should usually be cleared of their denominators through multiplication before you proceed to other equation-solving steps:

$$\frac{4 \times 3X}{4} = 6 \times 4$$

Simplify:

$$3X = 24$$

Now, divide each side by 3:

$$\frac{3X}{3} = \frac{24}{3}$$

Simplify:

$$X = 8$$

Check:

$$\frac{3X}{4} = 6$$

$$\frac{3 \times 8}{4} = 6$$

$$\frac{24}{4} = 6$$

$$6 = 6$$

EXAMPLE 3

$$\frac{27}{X} = 3$$

Solution:
First, multiply each side of the equation by X:

$$\frac{X \times 27}{X} = 3X$$

Simplify:

$$27 = 3X$$

You will notice that by multiplying each side of the equation by X, the X in the denominator, on the left-hand side of the equation, has been eliminated, and a "new" X has appeared on the right-hand side of the equation.

Next, divide each side of the equation by 3:

$$\frac{27}{3} = \frac{3X}{3}$$

Simplify:

$$9 = X \text{ or } X = 9$$

Check:

$$\frac{27}{X} = 3$$

$$\frac{27}{9} = 3$$

$$3 = 3$$

EXAMPLE 4

$$5X^2 + 6 = 51$$

Solution:
First, subtract 6 from each side of the equation:

$$5X^2 + 6 - 6 = 51 - 6$$

Simplify:

$$5X^2 = 45$$

Now, divide each side of the equation by 5; remember you are trying to isolate the variable:

$$\frac{5X^2}{5} = \frac{45}{5}$$

Simplify:

$$X^2 = 9$$

Finally, take the square root of each side:

$$\sqrt{X^2} = \sqrt{9}$$

Simplify:

$$X = 3$$

Check:

$$5X^2 + 6 = 51$$
$$5 \times 3^2 + 6 = 51$$
$$5 \times 9 + 6 = 51$$
$$45 + 6 = 51$$
$$51 = 51$$

EXAMPLE 5

$$8X + 4 = 3X - 6.$$

Solution:
First, combine the like terms (X) by subtracting $3X$ from each side:

$$8X + 4 - 3X = 3X - 6 - 3X$$

Simplify:

$$5X + 4 = -6$$

Now, subtract 4 from each side:

$$5X + 4 - 4 = -6 - 4$$

Simplify:

$$5X = -10$$

And to find the solution, divide each side by 5:

$$\frac{5X}{5} = \frac{-10}{5}$$

Simplify:

$$X = -2$$

Check:

$$8X + 4 = 3X - 6$$
$$8(-2) + 4 = 3(-2) - 6$$
$$-16 + 4 = -6 - 6$$
$$-12 = -12$$

EXAMPLE 6

$$\frac{4X - 9}{3} = 9$$

Solution:

First, multiply each side of the equation by 3 to clear the denominator:

$$\frac{3 \times (4X - 9)}{3} = 9 \times 3$$

Simplify:

$$4X - 9 = 27$$

Next, add 9 to each side of the equation to eliminate the loose constant on the left side:

$$4X - 9 + 9 = 27 + 9$$

Simplify:

$$4X = 36$$

And, to finally isolate X, divide each side of the equation by 4:

$$\frac{4X}{4} = \frac{36}{4}$$

Simplify:

$$X = 9$$

Check:

$$\frac{4X - 9}{3} = 9$$
$$\frac{4(9) - 9}{3} = 9$$
$$\frac{36 - 9}{3} = 9$$
$$\frac{27}{3} = 9$$
$$9 = 9$$

EXAMPLE 7

$$5(X + 3) = -35$$

Solution:

First, since X is contained in parentheses, use the distributive law to remove the parentheses. In other words, multiply each term within the parentheses by 5. In so doing, you are simplifying the left-hand side of the equation. The right-hand side of the equation is unaffected by this operation—you do not multiply the right-hand side by 5:

$$(5 \times X) + (5 \times 3) = -35$$

Now you can subtract 15 from each side of the equation:

$$5X + 15 - 15 = -35 - 15$$

Simplify:

$$5X = -50$$

The final step is to divide each side of the equation by 5:

$$\frac{5X}{5} = \frac{-50}{5}$$

Simplify:

$$X = -10$$

Check:

$$5(X + 3) = -35$$
$$5(-10 + 3) = -35$$
$$5(-7) = -35$$
$$-35 = -35$$

By now you have noticed that calculator solutions have not been shown for these problems, and for good reason . . . a variable cannot be entered into a simple calculator. Therefore, the problems must be solved in a step-by-step method, working toward isolating the unknown variable. Naturally, the calculator may be used to expedite the intermediary steps in the mathematical computations. Also, you should always check your work. Here again, the calculator can be an extremely useful tool.

▨ FORMULAS

A **formula** is simply an equation that states a well-known rule or law. Ohm's law, for example, is a formula that states that the current I is the ratio of the voltage V to the resistance R. $I = V/R$ The following examples are based on several well-known formulas, including the resistance of a wire, $R = (K \times L)/A$; slope of a straight line, $y = mx + b$; area of a triangle, $A = BH/2$; volume of a cylinder, $v = \pi r^2 h$; Celsius-to-Fahrenheit conversion, $°C = (5/9)(°F - 32)$; and time, rate, and distance, $d = r \times t$.

EXAMPLE 1

$$R = \frac{K \times L}{A}$$

Solution:
This formula already has R isolated, so it is in the proper form to determine the value of R.

Now, write an equivalent equation to solve for A:

$$R = \frac{K \times L}{A}$$

$A \times R = K \times L$ (multiply both sides by A)

$A = \dfrac{K \times L}{R}$ (divide both sides by R)

EXAMPLE 2

Using the initial formula given in Example 1, write an equivalent equation that will solve for L:

$$R = \frac{K \times L}{A}$$

$$A \times R = K \times L$$

$$L = \frac{A \times R}{K}$$

EXAMPLE 3

Now, take that same initial formula that you used in Example 2 and write an equivalent equation to determine K:

$$R = \frac{K \times L}{A}$$

$$R \times A = K \times L$$

$$K = \frac{R \times A}{L}$$

EXAMPLE 4

Consider the formula $y = mx + b$.
Rewrite the formula to solve for x:

$$y = mx + b$$

$$y - b = mx$$

$$x = \frac{y - b}{m}$$

EXAMPLE 5

If $A = \dfrac{BH}{2}$, what does H equal?

$$A = \frac{BH}{2}$$

$$2A = BH$$

$$H = \frac{2A}{B}$$

EXAMPLE 6

Given the formula $v = \pi r^2 h$. solve for h. (Note that pi, π, is used to represent a mathematical constant that is approximately equal to 3.14159.) Push the π key on your calculator and compare the display readout to the given pi value. A calculator maintains this value in its memory. Consequently, the mathematical value for pi is literally at your fingertips whenever you need it for a computation:

$$v = \pi r^2\, h$$

$$h = \frac{v}{\pi r^2} \qquad \text{(divide both sides by } \pi r^2)$$

Note in the preceding formula that both sides of the equation were divided by πr^2 in one step instead of dividing each side of the equation by π and then by r^2. Of course, this procedure could have been done in two steps, but it would have taken longer, and the result would be the same. Remember your basic rules. As long as each side of an equation is divided by the same value, it is a valid operational step.

EXAMPLE 7

Given the formula $^\circ C = \dfrac{5(^\circ F - 32)}{9}$, what does $^\circ F$ equal?

$$^\circ C = \frac{5(^\circ F - 32)}{9}$$

$$9\,^\circ C = 5(^\circ F - 32) \quad \text{(parentheses are no longer needed)}$$

$$\frac{9\,^\circ C}{5} = \,^\circ F - 32$$

$$\frac{9\,^\circ C}{5} + 32 = \,^\circ F$$

or

$$^\circ F = \frac{9\,^\circ C}{5} + 32$$

Once the variable has been isolated, its value can be determined on the basis of the other known quantities. Obviously, the next step involves simple substitution—taking the known quantities and substituting them into the given formula. Finally, the actual mathematical computations are performed to solve for the unknown variable.

EXAMPLE 8

The formula to solve for distance is $d = r \times t$, where d is the distance, r is the rate, and t is the time.

Using this formula, determine the distance that would be traveled at the rate of 40 miles per hour for 2½ hours:

$$d = r \times t$$

Given:

$r = 40$ miles per hour
$t = 2½$ hours

Solution:
Substitute the known values into the formula (the unknown is already isolated):

$$d = \frac{40 \text{ miles}}{1 \text{ hour}} \times 2.5 \text{ hours}$$

Perform the necessary mathematical computations (operations):

$$d = 100 \text{ miles}$$

Notice that when the problem is set up correctly, the hours units cancel each other, leaving the proper unit of miles in the answer.

EXAMPLE 9

Determine what your rate of travel is if you cover a distance of 500 miles in 4 hours. Use the distance formula found in the previous problem and rewrite it to solve for r:

$$d = r \times t$$
$$\frac{d}{t} = r \text{ or } r = \frac{d}{t}$$

Given:

$d = 500$ miles (mi)
$t = 4$ hours (hr)

Solution:

Substitute, using the given values:

$$r = \frac{500 \text{ mi}}{4 \text{ hr}}$$

Perform the required computation to solve the problem:

$$r = 125 \text{ mi/hr} = 125 \text{ mph}$$

EXAMPLE 10

How long would it take to travel 770 miles if you were moving at a rate of 55 mph?

Think of the distance formula and write an equivalent equation that will solve for time:

$$d = r \times t$$

$$\frac{d}{r} = t \text{ or } t = \frac{d}{r}$$

Given:

$$d = 770 \text{ mi}$$

$$r = 55 \text{ mph}$$

Solution:

Substitute using the known values:

$$t = \frac{770 \text{ mi}}{55 \text{ mi/hr}}$$

Perform the necessary computation:

$$t = 770 \text{ mi} \times 1 \text{ hr}/55 \text{mi}$$

Remember: When dividing by a fraction, invert the fraction and multiply:

$$t = 14 \times 1 \text{ hr}$$

$$t = 14 \text{ hr}$$

▓ WORD PROBLEMS

EXAMPLE 1

The length of a rug is 3 times its width. If the rug is 12 feet long, how wide is it? Choose appropriate letters to identify the variables. Let L = length and W = width.

Solution:

$$L = 3W$$

Isolate the required unknown variable (W in this case):

$$\frac{L}{3} = W \text{ or } W = \frac{L}{3}$$

Substitute for the known value of L and solve the equation:

$$W = \frac{12 \text{ ft}}{3}$$
$$W = 4 \text{ ft}$$

Check:

$$L = 3W$$
$$12 \text{ ft} = 3 \times 4 \text{ ft}$$
$$12 \text{ ft} = 12 \text{ ft}$$

EXAMPLE 2

Six years from now, Bob will be 3 times as old as he was 12 years ago. How old is Bob now? Logically thinking, let time in the future be represented by a positive value and time in the past by a negative value:

Solution:

Let X = Bob's age now.
$X + 6$ is Bob's age 6 years from now.
$X - 12$ is Bob's age 12 years ago.

Now, write a proper equation based on the given information:

$$X + 6 = 3(X - 12)$$

Solve for the unknown (X):

$$X + 6 = 3X - 36$$
$$6 + X - X = 3X - 36 - X$$
$$6 = 2X - 36$$
$$6 + 36 = 2X - 36 + 36$$
$$\frac{42}{2} = \frac{2X}{2}$$
$$21 = X \text{ or } X = 21$$

Answer: Right now, Bob is 21 years old.

Check:

$$X + 6 = 3(X - 12)$$
$$21 + 6 = 3(21 - 12)$$
$$27 = 3(9)$$
$$27 = 27$$

EXAMPLE 3

A handful of coins is made up of quarters and nickels. The coins have a total value of $1.35. There are 4 times as many nickels as there are quarters. Determine exactly how many quarters and nickels there are.

Logically thinking, in this problem it would be beneficial to think of the actual value of the coins in terms of cents. That is, a quarter represents 25 cents, and a nickel represents 5 cents. The total value of the coins involved could then be represented by the number of coins times the respective value. For example, 3 quarters would be represented as 3 times 25, or 75 cents.

Solution:

Let X = the actual number of quarters.

Let $25(X)$ = the total value of the quarters.

There are 4 times as many nickels as there are quarters.

Therefore, the number of nickels is represented by the term $4X$.

The value of the nickels involved would be $5(4X)$.

An equation can now be written to represent the value of all the coins involved:

$$25(X) + 5(4X) = 135 \text{ (cents in \$1.35)}$$

Solve the equation for X, which will determine the number of quarters:

$$25X + 20X = 135$$
$$45X = 135$$
$$X = 3 \text{ or 3 quarters}$$

The number of nickels involved is $4X$, or 4×3, which equals 12. Final answer: There are 3 quarters and 12 nickels.

Check:

$$25(3) + 5(4 \times 3) = 135$$
$$75 + 5(12) = 135$$
$$75 + 60 = 135$$
$$135 = 135$$

EXAMPLE 4

The area of a triangle is equal to ½ the product of the base times the height. If the base is 6 feet in length and the height is 9 feet, what is the area of the triangle?

Solution:
First, we need to translate the sentence into an equation. Let A = area, b = base, and h = height:

$$A = \frac{1}{2}bh$$

Now, substitute for the known quantities:

$$A = \frac{1}{2}(6 \text{ ft} \times 9 \text{ ft})$$

Solve the equation for A by performing the necessary computation:

$$A = \frac{1}{2} \times 54 \text{ ft}^2$$
$$A = 27 \text{ ft}^2$$

Check:

$$27 \text{ ft}^2 = \frac{1}{2}(6 \text{ ft} \times 9 \text{ ft})$$

$$27 \text{ ft}^2 = \frac{1}{2} \times 54 \text{ ft}^2$$

$$27 \text{ ft}^2 = 27 \text{ ft}^2$$

EXAMPLE 5

A certain unknown number is equal to the product of two other numbers divided by the sum of the same two other numbers. If the two other numbers are 8 and 2, what is the unknown number?

Solution:
Translate the preceding statements into an equation. Let X = the unknown number. Let A and B represent the two known numbers. In other words, let A = 8 and B = 2.
  Write the equation:

$$X = \frac{A \times B}{A + B}$$

Now, substitute for the known quantities:

$$X = \frac{8 \times 2}{8 + 2}$$

Solve for X:

$$X = \frac{16}{10} = 1.6$$

Check:

$$X = \frac{A \times B}{A + B}$$

$$1.6 = \frac{8 \times 2}{8 + 2}$$

$$1.6 = \frac{16}{10}$$

$$1.6 = 1.6$$

■ SUMMARY

In this chapter you learned one of the most important math skills that an electrician can have—the ability to manipulate and solve equations. Troubleshooting, designing, repairing, and even installing activities may require the calculation of a solution.

The steps are straightforward. First you must write the equation if it is not already given. This may be done from the statement of the problem or from your analysis of what needs to be done. In some cases the equation is developed from a known formula such as Ohm's law or the formula for the resistance of a wire.

The next step is to isolate the unknown variable on one side with all other known values on the other side of the equation. This is accomplished by performing inverse operations on both sides of the equation. For example, if 6 is added to one side of the equation, you move it to the other side by subtracting 6 from both sides. Balancing an equation always requires that you do exactly the same thing to both sides.

■ REVIEW QUESTIONS

1. What is the difference between a formula and an equation?

2. What is an equivalent equation, and how is it found?

3. What is the inverse action of multiplication? Of raising a number to a power?

4. A certain resistor is listed as 1,000,000 ohms. How many Megohms is this resistor? How many kilo-ohms?

5. Is $2 + 3 = 4 + 1$ an equation? Why or why not?

6. Is $2 + 3 = 4 - 1$ an equation? Why or why not?

7. What do an equation and a balance beam have in common?

8. In $5 = 3 + x$, which is the unknown variable?

■ PRACTICE PROBLEMS

1. Write a proper equation for the following statements.

   a. 7 plus 5 times some number is equal to 52.

   b. 6 less than some number squared is equal to 43.

   c. ¼ the product of a certain number times 8 is equal to 20.

   d. To find °C, subtract 32 from °F and divide this answer by 1.8.

e. To find °F, multiply °C by 1.8 and add 32 to this product.

2. Solve the following equations and check your answers by substituting the determined values for the variables in the original equations. These equations are fairly simple and provide excellent practice for developing the habit of always showing all the steps involved in finding a solution. This practice will go a long way toward eliminating inadvertent mistakes. As equations become more complex, the step-by-step approach will also make it easier to understand solutions and easier to determine them.

a. $X + 6 = 19$

b. $5 + Y = 4$

c. $X - 18 = 27$

d. $15 - X = 3$

e. $12X = 108$

f. $-6X = 36$

g. $X/6 = 7$

h. $-X/5 = 20$

i. $A2 = 25$

j. $C = 9$

3. Solve the following equations for the unknown variables. Check your answers and use your calculator where appropriate.

a. $9A + 6 = 42$

b. $68 = 12B - 4$

c. $\dfrac{8C}{3} = 24$

d. $\dfrac{42}{X} = -6$

e. $9X - 3 = 5X + 17$

f. $\dfrac{5R - 4}{3} = 17$

g. $4(3X - 5) = 28$

h. $\dfrac{5F}{4} - 7 = 8$

4. Using the following formulas, solve for the required unknown variable.

a. $P = IE$; solve for E

b. $E = mc^2$; solve for c

c. $P = 2A + 2B$; solve for B

d. $°F = \dfrac{9°C}{5} + 32$; solve for $°C$

e. $C^2 = A^2 + B^2$; solve for A

f. $V = a + gt$; solve for g

g. What is the value of B in question 3 if $P = 48$ and $A = 16$?

h. What is the value of °C in question 4 if °F = 212?

5. Solve the following word problems and check your answers. Take your time and walk yourself through them step by step.

a. Five less than twice some number is equal to one more than that number. What is the number?

b. If you have 2 dollars in change, made up of dimes and nickels, and you have 7 more nickels than you do dimes; how many dimes do you have?

c. The perimeter of a rectangle is equal to twice the length plus twice the width. If the perimeter of a certain rectangle is 44 feet and the length is 13 feet, what is the width?

d. Ten years from now, Jane will be three times as old as she was four years ago. How old is Jane now?

chapter 3

Using Ohm's Law and Associated Electrical Units

■ OUTLINE

■ OVERVIEW

In 1827, a German named Georg Simon Ohm published a formula that expresses the single most important relationship in all of electricity. This formula, named *Ohm's law* in honor of its discoverer, finds application in virtually all aspects of the electrical and electronic industries.

In this chapter you will learn how to apply Ohm's law in your day-to-day electrical responsibilities. As you work through the chapter, remember that although the examples are for DC circuits, Ohm's law will find application in AC as well. In fact, Ohm's law probably will be the most commonly used formula of all the ones that you learn and use as an electrician.

■ OBJECTIVES

After completing this chapter, you should be able to:

1. Describe the units of measurement of current, voltage, and resistance.
2. Demonstrate your knowledge of the units ohms, amperes, and volts by giving examples of their usage.
3. Explain the electron theory of current flow.
4. Calculate the resistance of wires given the length, cross-sectional area, and resistivity.
5. Solve electrical problems using Ohm's law.

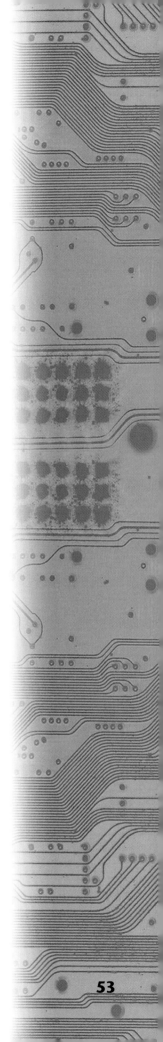

■ GLOSSARY

Ampere The unit of electrical current flow, often abbreviated as amp or just A.

ANSI American National Standards Institute, an American technical society that develops standards for electrical measurements and systems. ANSI standards often use English units, such as feet, inches, and pounds.

Conventional current flow The original definition of current direction in which current moves in the direction from a positive charge to a negative charge. Also called positive current flow and hole flow.

Coulomb The unit of electrical charge, equal to the total charge possessed by 6.25×10^{18} electrons. Abbreviated by the symbol C.

Current The motion of electrons through a material. Measured in amperes.

Electrical potential Electrical pressure created between a region of positive charge and a region of negative charge. Measured in volts.

Electromotive force Electrical pressure created between a region of positive charge (fewer electrons) and a region of negative charge (more electrons). Measured in volts.

Electron current flow The correct convention that states that current flows in the same direction that the electrons move through a conductor. This is the opposite convention to that used in so-called conventional current flow.

Hole flow The original definition of current direction in which current moves in the direction from a positive charge to a negative charge. Also called positive current flow and conventional current flow.

IEC International Electrotechnical Commission, a European technical society that develops standards for electrical measurements and systems. IEC standards often use SI units, such as meters, centimeters, and kilograms.

Joule The unit of electrical energy. One joule of energy used each second is equal to 1 watt. One joule is the amount of energy used when 0.737 pounds is lifted a distance of 1 foot.

kcmil Modern symbol for 1,000 circular mils. It has replaced the older MCM.

MCM Obsolete symbol for 1,000 circular mils (see kcmil).

Meter The unit of distance in the MKS system. One meter is equal to approximately 3.28 feet, or slightly over 1 yard.

Mil 1/1,000 of an inch.

Newton The unit of force in the MKS system. One Newton is equal to approximately 0.225 pounds.

Ohm The unit of electrical resistance, often shown as the Greek letter omega (Ω).

Ohm's law The mathematical relationship among current, electrical potential, and ohms when they are measured in amperes, volts, and ohms, respectively.

Positive current flow The original definition of current direction in which current moves in the direction from a positive charge to a negative charge. Also called conventional current flow and hole flow.

Resistance The physical opposition to electrical current. Resistance is caused by the energy loss that occurs when an electron displaces other electrons in a valence ring. Measured in ohms.

Resistivity The resistance of a conductor of a given unit length and a given unit cross-sectional area.

Specific resistance The same as resistivity.

Volt The unit of electrical pressure (electromotive force).

Voltage See Electromotive force.

Watt The unit of electrical power. One horsepower is equal to 745.7 watts.

■ INTRODUCTION

3.1 Historical Information

In your study of electricity, there are many key terms that you should learn. Three of these key terms are **current**, **voltage**, and **resistance**. In this chapter, these and other terms will be described in detail, along with an explanation of how they are used and how they relate mathematically to each other.

This mathematical relationship among current, voltage, and resistance is known as **Ohm's law**. Many of you have probably heard of Ohm's law either in high school, in technical school, or from some on-the-job experience.

3.2 Ohm's Law

Georg Simon Ohm was born on March 16, 1789, in Erlangen, Germany, and educated at the University of Erlangen (see Figure 3–1). From 1833 to 1849, he was director of the Polytechnic Institute of Nürnberg, and from 1852 until his death in 1854, he was professor of experimental physics at the University of Munich. Between 1825 and 1827, he developed a mathematical description of electrical current in circuits. What is now known as Ohm's law appeared in print in 1827. This work strongly influenced the electrical theory development of his day but was not well received by his peers.

Remember that when Ohm was doing his greatest work, there were no computers, no calculators, and no electric lights! He could use only earlier published manuscripts on physics, a slide rule, and his mind. Without the aid of modern technology, he published Ohm's law, which describes in detail the relationship among voltage, current, and resistance. Equation 1 is its simplified form, where I is the current of the circuit, E is the voltage applied to the circuit, and R is the total resistance of the circuit. The following sections discuss specific definitions relating to each term. The first sets of terms to be defined relate to current:

$$I = \frac{E}{R} \tag{1}$$

FIGURE 3–1   Georg Simon Ohm.

■ CURRENT FLOW

3.3 The Coulomb

In chapter 1, you learned that all atoms have electrons. These electrons travel in orbits around the atom's core, or nucleus, and each has a specific negative charge. The charge on each electron is essentially the same—all have the same amount of negative charge. This means that a group of 100 electrons have a charge 100 times larger than one electron. This was proved by a French physicist named Charles A. De Coulomb, who studied the movement of electrons through materials (see Figure 3–2). We call this flow of electrons through a material *current.* He proved that like charges repel (positive to positive) and that unlike charges (positive to negative) attract. He also proved that this repelling

FIGURE 3–2   Charles Coulomb.

force changes value as you change the distance between the charged materials.

Coulomb studied the amount of charge on an electron and the amount of force between them and based on his research defined the **coulomb** as equal to the total charge exhibited by 6.25×10^{18} electrons.

3.4 The Ampere

How do we use the definition of the coulomb? Another French scientist, Andre M. Ampere, used the coulomb in the definition of the **ampere** (see Figure 3–3). Ampere studied the effects of electrons flowing through wires in the early 1800s. He defined an ampere as 1 coulomb of electrons flowing past a specific point in a wire in 1 second.

While this definition is accurate, it was not consistent with the International System (SI) of weights and measurements that was adopted by the world in the twentieth century. The newer definition is defined based on the amount of force that is created between two wires carrying equal amounts of current.

You learned in chapter 1 that current flow creates a magnetic field. You also know that two magnetic fields will either attract or repel each other, depending on their polarity. The new definition of an ampere is based on these facts and is stated as follows:

When two long parallel wires, 1 meter apart, carry equal currents, and the magnetic force between the two wires is .0000002 (2×10^{-7}) newtons per meter, then the current flow in each wire is equal to 1 ampere.

3.5 Current Flow Convention

Conventional Current Flow

In the early days of electricity, before the development and understanding of electron theory, electricians did not know which way current flowed. They believed that something was flowing, and they needed to know which way it was flowing. Based on the magnetic field created and how it interacted with the Earth's magnetic field, early experimenters decided that current flowed from the positive pole to the negative pole of a DC source. This concept has been given at least three different names, including **hole flow**, **conventional current flow**, and **positive current flow**.

In modern times an explanation must be given to explain the concept of conventional current flow. When an electron leaves an atom, it creates a gap called a *hole* in the atom that it left. The atom with a missing electron is now a positive ion and will want to attract another free electron.

However, the new electron had to come from somewhere. This means that it made a hole somewhere else. If you look at the gaps' movements, it looks as if they are moving in the opposite direction of the electrons. Thus, we have a modern explanation for the concept of conventional current flow. Conventional current flow is still used in many engineering applications and in electrical engineering and physics classes in engineering colleges. Figure 3–4 illustrates the concept of conventional current flow.

FIGURE 3–3   Andre M. Ampere.

FIGURE 3–4 Hole flow concept used to explain "conventional" current flow.

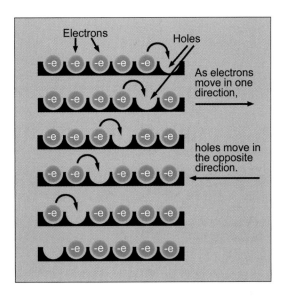

FIGURE 3–5 Current flow conventions.

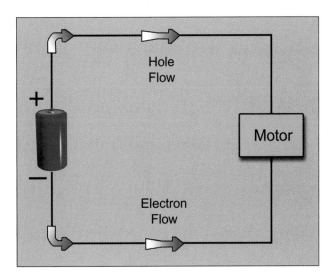

Electron Current Flow

Now, of course, we know that the electrons are actually moving through the wire. Ever since the mid-1900s, electrician and technician training has used the correct term, *electron flow,* in all training and day-to-day applications. This book and all the others you will use always use **electron current flow**. Figure 3–5 shows both conventions and their relative directions.

The maximum possible speed of the electrons is the speed of light in free space, which is equal to 299,792,458 meters per second (186,282 miles per second). In chapter 1, we learned that different materials allow electrons to move more freely than others; consequently, electrons will move more slowly in a conductor. There are many factors to take into account, including material, voltage , and frequency of the current flow. (You will learn about frequency when you study AC theory.) Generally, the current in a wire will flow at something less than the speed of light—often 80% of light speed or less.

■ THE VOLT

Using the definition of the coulomb, we can now define the **volt**. Voltage is the **electromotive force** (EMF) or a potential difference in electron charges. A potential difference exists when one object has more or less electrons than another object. Since each electron has a fixed amount of charge, there is potential energy available because of the two different electrical charges. This potential difference causes electrons to be repelled or attracted from one material to another (see Figure 3–6). In a circuit, this is from the negative side to the positive side of a battery. The volt is defined in terms of the coulomb and **joule**. (A joule is the amount of energy equal to .737 foot-pounds.)

The volt is defined as the amount of potential that will cause 1 coulomb to do 1 joule of work. Keep in mind it is the current doing the work, not the voltage. Voltage does not flow; it causes current flow, somewhat like a pump that causes water flow in a fluid system. The potential difference (EMF) in Figure 3–6 acts as a pump to force electrons to flow in the direction indicated.

■ THE OHM

Now that we've defined the ampere and the volt, let's look at the **ohm**. The ohm is a measurement of the amount of opposition to current flow that exists in an object. By definition, 1 ohm is the amount of resistance in a circuit that allows 1 ampere of current to flow when a potential of 1 volt is applied. This is a statement of Ohm's law in words. The symbol used for the ohm is the Greek symbol omega (Ω). Resistance is a property of all materials through which current can flow. Some materials allow current to flow better than others. If you recall from chapter 1, a material that has many free electrons is a conductor. Likewise, a material with few to no free electrons is an insulator. A conductor, therefore, has a low resistance, and an insulator has a high resistance.

■ TYPES OF CIRCUITS

3.6 Closed Circuit

All the knowledge you have can now be combined to create a circuit. You know that a potential difference (voltage) must exist to cause electrons to flow (current) within a conductor. And you know that all conductors will have some resistance (ohms) to that current flow.

There are many types of circuits you may encounter in your career as an electrical worker. Because of the ever changing designs for new electrical systems and circuits, not all are alike. You must become familiar with these differences to be successful. This section does not attempt to cover these circuits in detail but does give you an overview.

For current to flow, there must be a complete path, or a closed circuit (see Figure 3–7a). Notice that the electrons flow from the battery through the motor and back to the battery. The battery causes the potential difference for the electron flow. Another way to think of it is that the electrons make a solid chain through the circuit. If one electron moves, they all move in a circular path.

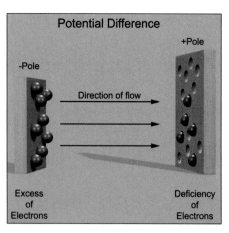

FIGURE 3–6   Electromotive force.

FIGURE 3-7   Three different types of circuits: (a) closed circuit; (b) open circuit ($R = \infty$); (c) short circuit ($R = 0$).

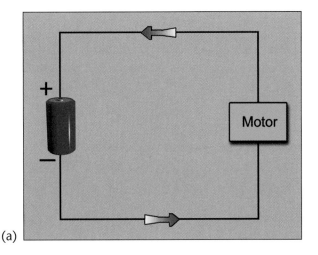

(a)

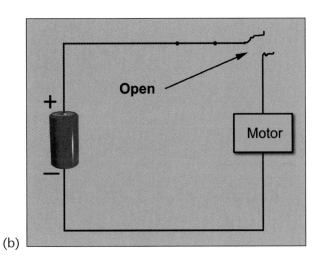

(b)

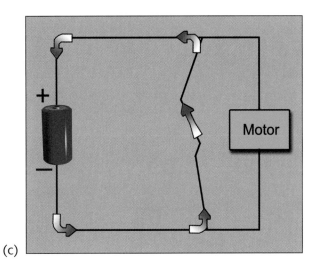

(c)

3.7 Open Circuit

If there is a break (open) somewhere in the circuit, the electrons cannot continue to flow (see Figure 3–7b). This is called an open circuit. The potential difference (voltage) still exists, but the electrons cannot move (there can be no current). This is another way of saying that R is infinite ($R = \infty$).

3.8 Short Circuit

If, in the path for current, there is a shorter and less resistive way for the electrons to get back to the source, most of the current (electrons) will take that path (see Figure 3–7c). This is called a *short circuit*. Short circuits can be very dangerous to anyone working on the system. A short in the circuit decreases the total resistance of the circuit because some higher-resistive circuit components have been bypassed. This causes the circuit current to increase (more electrons can flow because there is less resistance to their movement). A short circuit has almost zero resistance ($R = 0$).

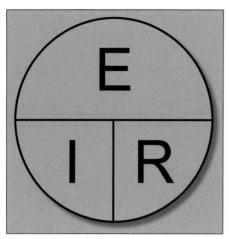

FIGURE 3–8   An Ohm's law pie.

■ CALCULATIONS INVOLVING OHM'S LAW

The relationship among current (I), voltage (E), and resistance (R) can be shown using a mathematical relationship called Ohm's law:

$$I = \frac{E}{R} \tag{2}$$

Other ways to write the relationship are in terms of E or R. The equations then look like this:

$$R = \frac{E}{I} \tag{3}$$

$$E = I \times R \tag{4}$$

An easy way to remember the relationship is to use the graphic representation in Figure 3–8. You can write any of these equations—just use your finger to cover up the value you are trying to find. Whatever remains is the right side of the equation.

EXAMPLE 1

If a circuit has a total resistance of 20 ohms (Ω) and a current of 3 amps, what is the circuit voltage?

Solution:
Using Figure 3–8, cover the "E" with your finger, and you have $I \times R$ left over. So the equation is

E = 3 × 20
E = 60 V

EXAMPLE 2

A certain circuit has an applied voltage of 24 volts and has a current of 2 amps. What is the resistance of the circuit?

Solution:
Again, looking at Figure 3–8, cover the "R," and the equation becomes

$$R = \frac{E}{I} \qquad R = \frac{24}{2} \qquad R = 12$$

■ ASPECTS OF CONDUCTORS

In most circuits, the voltage sources (e.g., batteries or generators) and loads (e.g., lamps or motors) are connected together using conductors in the form of wires. Wires are made of highly conductive materials, such as copper, and are generally coated or wrapped with insulating materials, such as rubber or plastic, to prevent the wires from touching and causing short circuits.

The amount of resistance exhibited by any given conductor is a function of four basic characteristics: cross-sectional area, type of material, length of conductor, and temperature of the conductor. The following sections describe the relationship of each of these characteristics to the resistance and provide convenient tables and some formulas that you can use for calculating the resistance of various wire sizes.

3.9 Cross-Sectional Area Effects on Conductor Resistance

Cross-Sectional Area of Conductors—ANSI

The cross-sectional area of American conductors is given in *circular mils* (see Figure 3–9). One mil is equal to .001 inch. The area of the conductor is calculated as the square of the diameter. For example, if the diameter of a given conductor is equal to 64.11 mils (.06411 inch), the cross-sectional area is equal to 64.11^2, which is equal to 4,110 circular mils (CM). For larger wire sizes, it is more convenient to express the cross-sectional area in thousands of circular mils. This measurement used to be called an **MCM**. Modern usage refers to it as a **kcmil**.

Table 3–1 shows that American wire sizes up above 4/0 are actually given in kcmils; 4/0 and below wires are given in American Wire Gauge sizes, so cross-sectional areas must be looked up or remembered.

To calculate the cross-sectional area of a conductor in CM, you measure the diameter of the conductor in inches, divide the result by 1,000, and square it.

FIGURE 3–9   A pictorial definition of circular mils (CM).

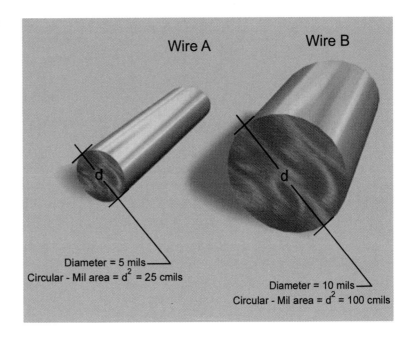

Wire A

Wire B

Diameter = 5 mils
Circular - Mil area = d^2 = 25 cmils

Diameter = 10 mils
Circular - Mil area = d^2 = 100 cmils

Table 3–1 Cross-Sectional Areas in Circular Mils for Various American Standard Wire Sizes

Size (AWG)	Area (Circular Mils)	Size (AWG)	Area (Circular Mils)	Size (AWG)	Area (Circular Mils)
18	1,620	1	83,690	600	600,000
16	2,580	1/0	105,600	700	700,000
14	4,110	2/0	133,100	750	750,000
12	6,530	3/0	167,800	800	800,000
10	10,380	4/0	211,600	900	900,000
8	16,510	250	250,000	1,000	1,000,000
6	26,244	300	300,000	1,250	1,250,000
4	41,740	350	350,000	1,500	1,500,000
3	52,620	400	400,000	1,750	1,750,000
2	66,360	500	500,000	2,000	2,000,000

EXAMPLE 1

You measure a conductor diameter and find that it is 0.5 inch. What is the cross-sectional area of the conductor?

$$A_{cm} = (d_{inches} \times 1,000)^2 = (0.5 \times 1,000)^2 = 250,000 \text{ CM}$$

Notice that 250,000 CM = 250 kcmils.

Cross-Sectional Area of Conductors—IEC

As the world becomes smaller, American electricians will see more international units of measure being used. Internationally, most conductor areas are expressed in square millimeters (mm²). In this system, the area is calculated using the familiar formula $A = \pi r^2$, where r is the radius in millimeters and $\pi = 3.14159$. By way of comparison, a wire that is 150 mm² is almost 300,000 cm, or 300 kcmil. The cross-sectional areas of IEC conductors are actually given in square millimeters, such as 100, 150, or 200.

To calculate the cross-sectional area of a conductor in mm², you measure the diameter in mm, divide it by 2, square it, and multiply the result by π.

EXAMPLE 2

You measure a conductor using a millimeter scale and find that the conductor diameter is very close to 16 mm. What is the cross-sectional area in mm²? (Note that if d is the diameter, then the radius is $r = d/2$.)

$$A = \left(\frac{d}{2}\right)^2 \times \pi = \left(\frac{16}{2}\right)^2 \times 3.14159 = 201 \text{ mm}^2$$

This cable is probably a 200-mm^2 cable since that is a standard metric size.

Variation of Resistance with Area

The resistance of a conductor is inversely proportional to its area. This is because a larger cross-sectional area will have more valence electrons available to participate in the current flow. Therefore, a 500-kcmil cable that is 1,000 feet long will have half as much resistance as the same length of a 250-kcmil cable.

3.10 Material Type Effects on Conductor Resistance

In your study of the structure of atoms, you found that each atom has a different atomic composition, and one characteristic of each atom is its unique electron configuration. Because of the varying atomic structures of different materials classified as conductors, each of these materials has a different resistance to current flow. In order to perform calculations to determine circuit resistance for a conductor, some information must be known about the particular properties of the material being used as the conductor. The necessary information is expressed in a term called **specific resistance**.

Specific resistance is also called **resistivity** and is usually abbreviated with the Greek letter ρ (rho) or by the capital letter K.

Specific resistance is defined for both American and IEC units:

ANSI: The resistance of a wire that is 1 foot long and 1 mil in diameter

IEC: The resistance of a wire that is 1 meter long and 1 meter in diameter[1]

While you could experimentally measure these values, they have already been determined for materials that may be used to conduct electricity. Table 3–2 lists the resistivity for many common materials in both the ANSI (American or English) and IEC (metric system).

Once the specific resistance or K value for a conductor material is known, then that value can be used in determining the overall resistance of any conductor made from the material. The relationship is such that the overall resistance of a conductor is directly proportional to the specific resistance of that conductor.

---

[1]Since the cross-sectional area of IEC standard wires are specified in mm^2, Table 3–2 shows metric values of ρ in terms of a wire that is 1 meter long and 1 millimeter in diameter, called 1 millimeter-meter.

Table 3–2   Resistivity (Specific Resistance) of Common Materials

Material	K (ρ) @ 68°F (20°C) American (English)	Metric	Resistance Temperature Coefficient (Ω/°)
Aluminum	17.7	.0265	.004308
Copper	10.4	.0168	.004041
Lead	126	.22	.0043
Mercury	590	0.98	.00088
Nichrome	600.0	1	.00017
Platinum	66	.106	.003729
Silver	9.7	.0159	.003819
Tungsten	33.8	.056	.004403

3.11 Material Length Effects on Conductor Resistance

For any given conductor, the resistance of that conductor is directly proportional to the length of the conductor. This means that the resistance for that conductor increases as the length increases. If a certain conductor has a resistance (R) for a given length of that conductor, then a conductor of the same diameter and material that is two times longer than the original length will have a resistance two times the original resistance or $2R$. Similarly, conductors 5 and 10 times longer will have resistance of $5R$ and $10R$.

3.12 Temperature Effects on Conductor Resistance

The specific resistance of any conductor is related to the temperature of that conductor; that is, the specific resistance changes as the temperature changes. Specifically, it increases if the coefficient is positive and decreases if it is negative. Most conductors have positive temperature coefficients, while most insulators have negative temperature coefficients. Carbon is one of the few materials that is classed as a conductor that has a negative temperature coefficient. Table 3–2 lists the specific resistances for the various materials when their temperature is 68°F (20°C).

EXAMPLE 3

Find the resistivity of copper at 90°F.

Solution:
Since Table 3–2 gives the temperature coefficient in Ω/°C, it will be easier if you first convert 90°F to degrees Celsius (°C). This can be

done with the formula $°C = \frac{5}{9}(°F - 32) = \frac{5}{9}(90 - 32) = 32.22°C.$

Since the coefficient for copper is .004041, the change in resistivity will be .004041 × (32.22 − 20) = 0.04938102.

Since the coefficient is positive, you add it to the coefficient at 20°C. Thus, the resistivity of copper at 90°F = .0168 + .04938102 = .06618.

3.13 Determining the Resistance of a Conductor

Once all information is gathered on a specific conductor's physical attributes, it becomes a simple task to determine the resistance for any length of that conductor. The resistance (R) in ohms is defined as being equal to the specific resistance (K or ρ) times the length (L) divided by the area (A). When written out the formula takes the form

$$R = \frac{K \times L}{A} \qquad (5)$$

Note that you must use the proper units for the type of conductor you are working with. For example, in the American system you must use the English value for K, L is in feet, and A is in circular mils. In the metric (IEC) system, use the metric value of K, L is in meters, and A is in square millimeters.

When the conductor material, length, and area are known for a conductor, the resistance is found by substituting values into the formula and then solving for the resistance.

The American Wire Guage (AWG) information in Table 3–3 shows characteristics of solid copper conductor up to 4/0 in size.

EXAMPLE 4

If a 6-gauge wire is made of copper and is 1,000 feet long, what is the resistance of the wire at 68°F?

$$R_{ohms} = \frac{10.4 \times 1,000}{26,244}$$

$$R_{ohms} = 0.396 \ \Omega$$

This same formula is used quite frequently in a different form for determining voltage drops in conductor calculations. Since resistance is an essential part of any electrical circuit, the ability to be able to calculate resistance in conductors is an important part of the knowledge required when working with them.

Table 3–3 American Wire Gauge Table

B&S Gauge No.	Diameter in Mils	Area in Circular Mils	Ohms per 100 Feet (Ohms per 100 m) Copper* 68°F (20°C)		Copper* 167°F (75°C)		Aluminum 68°F (20°C)		Pounds per 1,000 Feet (kg per 100 m) Copper		Aluminum	
0000	460	211,600	0.049	(0.016)	0.0596	(0.0195)	0.0804	(0.0263)	640	(95.2)	195	(29.0)
000	410	167,800	0.0618	(0.020)	0.0752	(0.0246)	0.101	(0.033)	508	(75.5)	154	(22.9)
00	365	133,100	0.078	(0.026)	0.0948	(0.031)	0.128	(0.042)	403	(59.9)	122	(18.1)
0	325	105,500	0.0983	(0.0983)	0.1195	(0.0392)	0.161	(0.053)	320	(47.6)	97	(14.4)
1	289	83,690	0.1239	(0.0406)	0.151	(0.049)	0.203	(0.066)	253	(37.6)	76.9	(11.4)
2	258	66,370	0.1563	(0.0512)	0.190	(0.062)	0.256	(0.084)	201	(29.9)	61.0	(9.07)
3	229	52,640	0.1970	(0.0646)	0.240	(0.079)	0.323	(0.106)	159	(23.6)	48.4	(7.20)
4	204	41,740	0.2485	(0.0815)	0.302	(0.099)	0.408	(0.134)	126	(18.7)	38.4	(5.71)
5	182	33,100	0.3133	(0.1027)	0.381	(0.125)	0.514	(0.168)	100	(14.9)	30.4	(4.52)
6	162	26,250	0.395	(0.1290)	0.481	(1.158)	0.648	(0.212)	79.5	(11.8)	24.1	(3.58)
7	144	20,820	0.498	(0.163)	0.606	(0.199)	0.817	(0.268)	63.0	(9.37)	19.1	(2.84)
8	128	16,510	0.628	(0.206)	0.764	(0.250)	1.03	(0.338)	50.0	(7.43)	15.2	(2.26)
9	114	13,090	0.792	(0.260)	0.963	(0.316)	1.30	(0.426)	39.6	(5.89)	12.0	(1.78)
10	102	10,380	0.999	(0.327)	1.215	(0.398)	1.64	(0.538)	31.4	(4.67)	9.55	(1.42)
11	91	8,234	1.260	(0.413)	1.532	(0.502)	2.07	(0.678)	24.9	(3.70)	7.57	(1.13)
12	81	6,530	1.588	(0.520)	1.931	(0.633)	2.61	(0.856)	19.8	(2.94)	6.00	(0.89)
13	72	5,178	2.003	(0.657)	2.44	(0.80)	3.29	(1.08)	15.7	(2.33)	4.8	(0.71)
14	64	4,107	2.525	(0.828)	3.07	(1.01)	4.14	(1.36)	12.4	(1.84)	3.8	(0.56)
15	57	3,257	3.184	(0.043)	3.98	(1.27)	5.22	(1.71)	9.86	(1.47)	3.0	(0.45)
16	51	2,583	4.016	(0.316)	4.88	(1.60)	6.59	(2.16)	7.82	(1.16)	2.4	(0.36)
17	45.3	2,048	5.06	(1.66)	6.16	(2.02)	8.31	(2.72)	6.20	(0.922)	1.9	(0.28)
18	40.3	1,624	6.39	(20.9)	7.77	(2.55)	10.5	(3.44)	4.92	(0.731)	1.5	(0.22)
19	35.9	1,288	8.05	(2.64)	9.79	(3.21)	13.2	(4.33)	3.90	(0.580)	1.2	(0.18)
20	32.0	1,022	10.15	(3.33)	12.35	(4.05)	16.7	(5.47)	3.09	(0.459)	0.94	(0.14)
21	28.5	810	12.8	(4.2)	15.6	(5.11)	21.0	(6.88)	2.45	(0.364)	0.745	(0.110)
22	25.4	642	16.1	(5.3)	19.6	(6.42)	26.5	(8.69)	1.95	(0.290)	0.591	(0.09)
23	22.6	510	20.4	(6.7)	24.8	(8.13)	33.4	(10.9)	1.54	(0.229)	0.468	(0.07)
24	20.1	404	25.7	(8.4)	31.2	(10.2)	42.1	(13.8)	1.22	(0.181)	0.371	(0.05)
25	17.9	320	32.4	(10.6)	39.4	(12.9)	53.1	(17.4)	0.97	(0.14)	0.295	(0.04)
26	15.9	254	40.8	(13.4)	49.6	(16.3)	67.0	(22.0)	0.77	(0.11)	0.234	(0.03)
27	14.2	202	51.5	(16.9)	62.6	(20.5)	84.4	(27.7)	0.61	(0.09)	0.185	(0.03)
28	12.6	160	64.9	(21.3)	78.9	(25.9)	106	(34.7)	0.48	(0.07)	0.147	(0.02)
29	11.3	126.7	81.8	(26.8)	99.5	(32.6)	134	(43.9)	0.384	(0.06)	0.117	(0.02)
30	10.0	100.5	103.2	(33.8)	125.5	(41.1)	169	(55.4)	0.304	(0.04)	0.092	(0.01)
31	8.93	79.7	130.1	(42.6)	158.2	(51.9)	213	(69.8)	0.241	(0.03)	0.073	(0.01)
32	7.95	63.2	164.1	(53.8)	199.5	(65.4)	269	(88.2)	0.191	(0.02)	0.058	(0.01)
33	7.08	50.1	207	(68)	252	(82.6)	339	(111)	0.152	(0.01)	0.046	(0.01)
34	6.31	39.8	261	(86)	317	(104)	428	(140)	0.120	(0.01)	0.037	(0.01)
35	5.62	31.5	329	(108)	400	(131)	540	(177)	0.095	(0.01)	0.029	
36	5.00	25.0	415	(136)	505	(165)	681	(223)	0.076	(0.01)	0.023	
37	4.45	19.8	523	(171)	636	(208)	858	(281)	0.0600	(0.01)	0.0182	
38	3.96	15.7	660	(216)	802	(263)	1080	(264)	0.0476	(0.01)	0.0145	
39	3.53	12.5	832	(273)	1012	(332)	1360	(446)	0.0377	(0.01)	0.0115	
40	3.15	9.9	1049	(344)	1276	(418)	1720	(564)	0.0299	(0.01)	0.0091	
41												
42	2.50	6.3										
43												
44	1.97	3.9										

*Resistance figures are given for standard annealed copper. For hard-drawn copper, add 2%.

■ SUMMARY

Energy is transferred from one part of an electrical circuit to another by causing current to flow from an energy source (such as a battery or generator) to a load (such as a light, heater, or motor). The flow of current has been shown to be the movement of electrons through the wire. The electrons move from atom to atom by replacing each other in the valence shell.

The three most fundamental quantities in electricity are the current (rate of electron flow), the voltage (electrical pressure that causes current flow), and the resistance (the physical opposition to current flow). You will spend a great deal of your career dealing with all these quantities. Remember the following:

- Voltage is required to select wire and other electrical devices that have the proper insulation rating.

- Current is required to know what size wire should be used for a given application or what size transformers or generators should be used.
- Resistance is used as the key element in determining voltage drop for long runs of cable.

Fortunately, there is a mathematical relationship among these three parameters that can be used to calculate any one of them when the other two are known. That formula is called Ohm's law and is simply stated as $I = E/R$, where I is the current measured in amperes, E is the electrical pressure in volts, and R is the electrical resistance in ohms.

Another useful formula is the one that relates the length, cross-sectional area, and resistivity characteristics for a particular conductor. This formula is also relatively simple and is given as $R = (\rho \times L)/A$.

■ REVIEW QUESTIONS

1. Write Ohm's law in its three forms.
2. A piece of 150-mm^2 copper wire is 3,000 meters long. What is its resistance at 20°C?
3. Describe the way in which resistance varies with conductor length, size, and temperature.
4. What does a positive resistance temperature coefficient mean? A negative coefficient?
5. Describe the concept of electron current flow.
6. What is conventional current flow (also called hole flow)?

7. What is a coulomb, and how does it relate to electric current?
8. Discuss the differences among closed circuits, open circuits, and short circuits.
9. The diameter of a certain wire is 0.5 inch (12.7 mm). What is its cross-sectional area in circular mils (CM)? In kcmils? In square millimeters?
10. What is the cross-sectional area of #8 wire in kcmils? What is the cross-sectional area of 350-kcmil wire in CM?

chapter 4

The Properties of Power in an Electrical Circuit

■ OUTLINE

■ OVERVIEW

The ultimate purpose of electricity used in power applications is to do work—in other words, electrical systems transfer energy from one place (a source) to another (a load). To do work requires the expenditure of energy; in fact, *work* and *energy* are interchangeable terms. Many of your tasks as an electrician will involve the calculations of power, which is simply work divided by time.

In this chapter you will learn the definitions (formulas) of work and how to manipulate them. You will also learn many different units of energy and power and how to convert from one to the other. The discussions in this chapter cover DC circuits only, although the concepts will apply in AC circuits as well. You will learn in AC circuits that the idea of power and energy is a little more sophisticated. For now, remember that all energy (work) in a DC system is either dissipated as heat or used to perform some physical effort, such as turning a pump shaft or lighting a lightbulb.

■ OBJECTIVES

After completing this chapter, you should be able to:

1. Describe how electrical power is utilized or dissipated.
2. Mathematically solve circuit problems using Ohm's law for power calculations.
3. Explain the units of measurement for both mechanical and electrical power.
4. Convert among the various units of mechanical and electrical power.
5. Calculate % efficiency given the watts supplied and the watts lost for a given load.

■ GLOSSARY

Energy   The ability to do work.
Power   Work done per unit time.

Work   The act of transferring energy from one place to another. The terms *work* and *energy* are often used interchangeably, although work is actually a change or expenditure of energy with no concern for time.

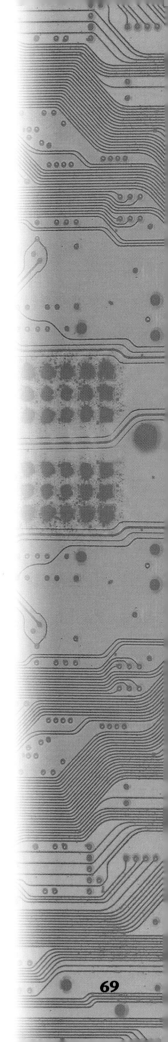

■ INTRODUCTION

In chapter 2, you studied Ohm's law and studied calculations. The units you used were the ohm, the ampere, and the volt. In this chapter, you will study the **power** aspects of an electrical circuit. Like the ohm, ampere, and volt, the amount of power can be specifically calculated. It would be helpful here to review the units used in Ohm's law because the electrical power consumed by an electrical circuit is calculated using those units.

■ REVIEW OF OHM'S LAW

4.1 Ampere

The ampere, or amp, is the measure of a specific number of electrons that pass a specific point in 1 second. That number is approximately 6.25×10^{18} electrons and is called a *coulomb*. When that number of electrons passes a specific point in 1 second, we say that 1 amp is flowing and is represented by the symbol A. In calculations we use the letter I.

4.2 Volt

The volt is a measure of electromotive force that pushes electrons through the wires and components of a circuit. It is similar to the pressure exerted on a system of fluid using pipes. The higher the pressure, the greater the flow. Specifically, the volt is the amount of electrical pressure between two points that will cause 1 coulomb of charge to do 1 joule of work. The volt is given the symbol V. In calculations we use the letter E. Remember that voltage is the force required in creating flow but that volts do not flow through the circuit.

4.3 Ohm

The ohm is the unit of resistance in a circuit. Specifically, it is the amount of resistance that allows 1 amp of current to flow when 1 volt is applied. The symbol used to represent the ohm is omega (Ω). In calculation we use the letter R. A component in a circuit that creates resistance is called a *resistor*.

4.4 Ohm's Law

Ohm's law is a law of proportionality that states that it takes 1 volt to push 1 amp through 1 ohm of resistance. Another way to look at it is that the current I in a circuit is directly proportional to the voltage E applied to the circuit and inversely proportional to the resistance R.

Figure 4–1 can help you remember the relationship. To find the value you are looking for, cover it with your finger to see the relationship that remains. From Figure 4–1, you can derive the three equations you need to make any calculation using Ohm's law:

$$E = IR \tag{1}$$

$$I = \frac{E}{R} \tag{2}$$

FIGURE 4–1 The Ohm's law pie chart

$$R = \frac{E}{I} \qquad (3)$$

4.5 Power

$$\text{Power} = \frac{\text{Work}}{\text{Time}} \qquad (4)$$

Another value that is helpful in designing and working on electrical circuits is the power requirement. In electricity, the unit of power is the watt. Power in a circuit is the amount of **work** being done per unit time (Equation 4). Power is what is consumed when a voltage (volts) is applied to a circuit and current (amps) flows through the load. The relationship can be seen between the applied voltage and the current consumed by the load as power is consumed. This is shown in Equation 5.

$$P = E \times I \qquad (5)$$

The amount of power consumed by a load is directly related to the amount of voltage applied to the load and the amount of current flowing through the load.

The load in the circuit uses the power to change or convert electrical **energy** into some other form of energy. For power to exist in an electrical circuit, this change must take place. The change may be in the form of current flow causing a heater element to heat up (change from electrical energy to heat) or a voltage supplied to a motor causing the motor to rotate (electrical energy to mechanical energy). The key point to remember is that the change or conversion from one form to another must take place.

The more current that is allowed to flow through a circuit, the hotter it gets. This heat is produced by the collision of the flowing free electrons with the fixed atoms. But that isn't the strange part: If the current is doubled, the heat produced will go up by four times, and if the current is tripled, the heat produced goes up by nine times! This showed early scientists that the heat produced, and hence the work being done in the circuit, was proportional to the force applied but not the current. The proportion was of an unexpected nonlinear nature.

English physicist James Prescott Joule (1818–1889) dealt with this problem of nonlinear proportionality. He wrote a paper in 1840 called "On the Production of Heat by Voltaic Electricity." In that paper, he explained his experiments, which we won't go into here. His conclusions are called Joule's law.

Joule's law states that "the total amount of heat produced in a conductor is directly proportional to the resistance times the square of the current" (see Figure 4–2). Mathematically, the heat, or power, produced can be shown in Equation 6:

$$P = I^2 \times R \qquad (6)$$

The heat produced by the current through the resistance in the circuit is called "I^2R" losses because it is heat lost in the system.

In Equation 6, the formula is found by combining the formula $P = IE$ with the formula $E = IR$. By substituting IR for E in the $P = IE$ formula ($P = I \times [I \times R]$), you get $P = I^2R$.

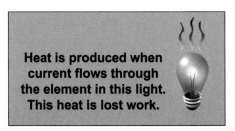

Heat is produced when current flows through the element in this light. This heat is lost work.

FIGURE 4–2   Heat and light are produced by the electric lightbulb.

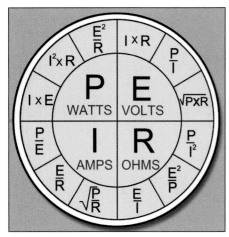

FIGURE 4–3 More detailed Ohm's law pie chart.

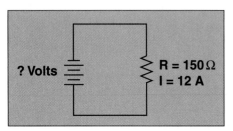

FIGURE 4–4 DC circuit for Example 1.

Because of Joule's work on electrical heat, he is credited with proving the relationship between electrical power and current and was given the honor of having a unit named after him: the joule. The joule is an amount or unit of energy used to measure the heat or work produced or consumed in a system. In an electrical circuit, this is the heat produced or the work done in the circuit due to the amount of current flowing through it. The work used in heating a circuit is called *lost work* since this work cannot be used to move anything.

Because of the heat generated in electrical components, specifically resistors, a rating system has been developed so components will not overheat during normal operation. This power rating is calculated using Equations 1, 2, 3, and 6. By substituting the different values for I and R in Equation 6, you come up with Equations 7 and 8:

$$P = \frac{E^2}{R} \tag{7}$$

$$P = I\mathrm{E} \tag{8}$$

Further, solving for each of the values I, E, and R, the following relationship diagram can be derived. Using Figure 4–3, you can solve most equations involving power, current, voltage, and resistance. The chart in Figure 4–3 is found by combining the formula E = IR with the formula $P = I$E. Other variations of the relationship between these variables can be found, such as $I = P/$E and E = P/I.

EXAMPLE 1

What is the voltage requirement for a circuit with resistance 150 Ω and a current of 12 A as shown in Figure 4–4?

Solution:
Using Figure 4–3, you can see that in the voltage quadrant, there are three formulas to choose from. Since you know resistance and current, the equation becomes

E = IR

E = 12 × 150

E = 1,800 V

EXAMPLE 2

What is the power lost due to heat in a motor that draws 20 A of current and has a resistance 5 Ω?

Solution:
In Figure 4–3, the power quadrant shows that since you know current and resistance, the formula to use is

$P = I^2R$

$P = 20^2 \times 5$

$P = 2,000$ W

EXAMPLE 3

A room heater has a rating of 2,000 W and uses 120 V to power it. What is the resistance of the heater?

Solution:
In Figure 4–3, the resistance quadrant shows that since you know the power and the voltage, the formula to use is

$$P = \frac{E^2}{R}$$

$$R = \frac{E^2}{P}$$

$$R = \frac{120^2}{2,000}$$

$$R = 7.2 \ \Omega$$

EXAMPLE 4

Using Figure 4–5, what is the power consumed by the lamp?

Solution:
In Figure 4–3, the power quadrant shows that since you know the voltage and the current, the formula that you use to solve for the circuit power is

$$P = I \times E$$

$$P = .5 \times 10$$

$$P = 5 \ W$$

EXAMPLE 5

Assume, based on the circuit in Figure 4–5, that the circuit power was 75 watts and the circuit voltage was 3 Volts. Calculate the circuit current.

Solution:
In Figure 4–3, the current quadrant shows that since you know the circuit power and circuit voltage, the formula that you use to solve for the circuit is

$$I = \frac{P}{E}$$

$$I = \frac{75}{3}$$

$$I = 25 \ A$$

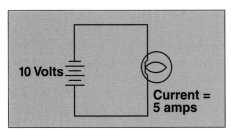

FIGURE 4–5   DC lamp circuit for Example 4.

EXAMPLE 6

Assume that the circuit in Figure 4–5 has a circuit current of 7 amps and consumes 42 watts of power. Calculate the circuit voltage.

Solution:
In Figure 4–3, the voltage quadrant shows that since you know the circuit current and the circuit power, the formula that you use to solve for the circuit voltage would be

$$E = \frac{P}{I}$$

$$E = \frac{42}{7}$$

$$E = 6 \text{ V}$$

■ INCANDESCENT LAMPS

Incandescent lamps show the relationships among current, voltage, resistance, and power. Lamps have both a voltage rating and a power rating. Common values used in homes are 120 V/100 W, 120 V/75 W, 120 V/60 W, and 120 V/40 W. These lamps are also available in other voltage and wattage ratings. Note that even though incandescent lamps are intended for use in AC circuits, they are rated in such a way that the power and current ratings are the same in DC circuits.

4.6 Ratings

Different wattage ratings are achieved for lamps with the same voltage ratings by changing the value of resistance. Higher-wattage bulbs will have a lower resistance and therefore a higher current. Typical resistances are 100-watt lamp, 144 ohms; 75-watt lamp, 192 ohms; and 40-watt lamp, 360 ohms.

EXAMPLE 1

What is the resistance of a 60-watt lamp?

Solution:
Using the previous information and from the resistance quadrant in Figure 4–3, the equation to use is

$$P = \frac{E^2}{R}$$

$$R = \frac{E^2}{P}$$

$$R = \frac{120^2}{60}$$

$$R = 240 \ \Omega$$

4.7 Lost Power

In an incandescent lamp, the current flowing through the filament causes the lamp filament to heat to incandescence, which means it emits visible light. Generally, the heat generated by the lamp can be classified as a loss since it is unwanted. Because the intended purpose of the lamp is to provide light, not heat, the power loss resulting from this unwanted heat is often identified as "I^2R" losses. However, the resistive components in a circuit are the only ones that will generate this heat. Only resistance can use power in an electrical circuit. Even the apparent resistance of an electric motor can be calculated; consequently, the power used to supply an electric motor can be calculated using these same formulas.

4.8 Efficiency

Efficiency has become a very important concept in our modern world. Efficiency is a measure of how much energy (or power) is being used for useful work, such as light, and how much is being wasted. If a 75-watt incandescent lamp uses 55 watts to generate light and the other 20 watts is dissipated as heat, then the % efficiency for the lamp would be

$$\% \text{ Efficiency} = \frac{(75 \text{ W} - 20 \text{ W})}{75 \text{ W}} \times 100 = 73.3\%$$

From this you can see that quite a large portion of the energy you pay for in your home goes into heating rooms, not lighting them.

■ OTHER USEFUL UNITS

There are many different types of power units, and you should be familiar with a few of them. You have covered one already: the watt. However, the watt is a relatively small number when compared to the voltage and current ratings of commercial electrical circuits and components. To make the numbers a little more manageable, the electrical commercial industry has adopted the use of the kilowatt, or thousand-watt, unit. One kilowatt is the same as 1,000 watts. An easy relation to this number can be observed in your own home. The rating for incandescent lightbulbs is in watts. A standard light might be 100 watts. So, you can see that it wouldn't take many lights to use 1 kilowatt: 10 to be exact.

The kilowatt is a common unit of electrical power, but it is more helpful for commercial electric companies to know how much energy has been used over a period of time, such as every month. Since the kilowatt is the amount of work being performed per unit time (second), we simply multiply by a unit of time (hour) and derive a new unit: the kilowatt-hour. This new unit is really just a different way of saying how much total energy has been used. Each month, the electric company in your area will read the kilowatt-hour meter on your home to let you know (and bill you for) how much energy your home has used.

Another unit is the horsepower. James Watt, who was a steam engine builder, decided that to make people want to buy his engines, he would have to give the engines a rating system the average person could understand. He found, through experimentation, that an average workhorse could lift 550 pounds 1 foot in 1 second. This was the amount of work being done by the horse in 1 second. This number is the rating he used for the horsepower. One horsepower is equal to 550 foot-pounds per second. This is still a standard today. He also found through calculations that 1 horsepower was equal to approximately 746 watts.

The calculation to convert horsepower to watts is easily accomplished.

EXAMPLE 1

A 3-horsepower motor draws how many watts?

Solution:
Using the previous information and knowing that 1 horsepower is equal to 746 watts, the conversion would use the following formula:

$$P = \text{hp} \times 746$$
$$P = 3 \times 746$$
$$P = 2{,}238 \text{ W, or } 2.238 \text{ kW}$$

EXAMPLE 2

A DC motor draw 3,730 watts. What would be the horsepower rating for this motor?

$$\text{hp} = \frac{P}{746}$$
$$\text{hp} = \frac{3{,}730}{746}$$
$$\text{hp} = 5$$

There are other units that you will become familiar with. The British Thermal Unit (BTU) is commonly used in the heating, air conditioning, and refrigeration industries. A BTU is the amount of energy required to raise the temperature of 1 pound of water 1 degree Fahrenheit. The metric system equivalent of the BTU is the calorie. The calorie is the amount of energy required to raise 1 gram of water 1 degree Celsius.

The International System (SI) of Weights and Measurements, which was adopted by the United States in October 1960, has six basic units: length, mass, time, temperature, electric current, and light intensity. Table 4–1 defines each of these.

The units of electrical measurements are defined in terms of the older physical quantities. The definitions in Table 4–2 will help you understand them.

Table 4–1 The Six Basic Units of the International System (SI) of Weights and Measurements

Length: meter	The meter is defined as 1,650,763.73 wavelengths in a vacuum of the orange-red line of the spectrum of krypton-86.
Mass: kilogram	This standard is an artifact of the mass weight of a cylinder of platinum-iridium alloy kept by the International Bureau of Weights and Measures in Sevres, France.
Time: second	This is the time required for 9,192,631,770 cycles of the radiation associated with specified transition of change in energy level of radiated cesium-133.
Temperature: Kelvin (K)	The Kelvin measure of temperature was adopted with the zero point defined at 273.16 K where ice, liquid water, and water vapor are in equilibrium. This is equivalent to .01°C and 32.02°F.
Electric current: ampere	This is the amount of electron flow through two long parallel wires separated by 1 meter in free space that results in a force between the wires for each meter of length of 2×10^{-7} newtons due to the magnetic field created by the current flow.
Light intensity: candela (cd)	This is the luminous intensity of 1/600,000 of a square meter of radiating cavity at the temperature of freezing platinum (2,042 K).

Table 4–2 Complex Electrical Units of Measurement as Developed from the SI Basic Units

Speed	Meters/second
Acceleration	Meters/second^2
Force	Mass $\times$ acceleration. When gravity is substituted for acceleration, force is equal to weight. The measure of weights is newtons (N) in the metric system and pounds (lb) in the English system.
Potential energy	The ability to do work. The unit is the joule in the metric system and is equal to about .737 foot-pounds in the English system. The volt is the electrical unit of potential energy.
Work	Force $\times$ distance. The joule is the unit of measure in the metric system, and the foot-pound is used in the English system. In order for work to be done, a weight must be moved a distance.
Power	The rate of doing work. The watt is the unit of electrical power.
Electrical energy use	Usage of electrical energy is measured in watt-hours or kilowatt-hours.
Horsepower	The unit of measurement of mechanical power. One horsepower is equal to 550 foot-pounds per second.

Table 4–3   Conversion Factors of Common Units

1 horsepower	746 watts
1 horsepower	550 ft-lb/sec
1 BTU/hr × 0.293	watts
1 cal/sec	4.19 watts
1 ft-lb/sec	1.36 watts
1 BTU	1,050 joules
1 joule	0.2389 cal
1 cal	4.186 joules
1 watt	0.00134 horsepower
1 watt	3.1412 BTU/hr
1 watt-second	1 joule

You should become familiar with Table 4–3. It gives the relationship between commonly used units, and you will find it helpful when performing calculations for different values.

■ SUMMARY

Electrical energy is transferred through wires from sources such as generators or batteries to loads such as lightbulbs, heaters, or motors. This energy is expended over time by providing light, heat, or motion. Unfortunately, some of the energy is wasted as heat (as in lightbulbs). The modern emphasis on energy conservation has created a great interest in calculating efficiencies with the idea of improving them so that less energy is wasted on a day-to-day basis.

Fortunately, the relationship among power, current, voltage, and resistance is easily calculated by using extensions of Ohm's law. As soon as any two values are known, the others can be calculated using these formulas.

Of course, any time that a formula is used, the units must be carefully chosen. Just as the basic Ohm's law works only when the voltage, current, and resistance are measured in volts, amperes, and ohms, respectively, so too the power versions of Ohm's law will work only when the power is measured in watts. If other units of power are required, such as horsepower, a conversion must be made.

■ REVIEW QUESTIONS

1. Explain how work and power are related to each other. Use the formula for power in your explanation.

2. Explain how work and energy are related to each other.

3. How would you calculate power if you know (a) the voltage and the current, (b) the current and the resistance, and (c) the voltage and the resistance?

4. Which would you rather use: a 40-watt lightbulb with 90% efficiency or a 60-watt lightbulb with 50% efficiency? Why?

5. The efficiency of a certain electrical circuit is 80%. At least some of the losses occur because of heating of the wire. How could the overall efficiency be improved?

6. Describe the meaning and relationship among voltage, current, and resistance.

7. Explain why all the power of a 100-watt incandescent lamp is not turned to light.

8. Explain the concept and the development of the unit of horsepower and how it relates to the watt or kilowatt.

9. An incandescent lamp uses heat to create light. A fluorescent lamp doesn't use heat. Generally, which would you expect to be the most efficient? Why?

10. Refer to Figure 4–3 and discuss the method(s) that you might use to calculate any one value when two others are known.

■ PRACTICE PROBLEMS

1. What is the resistance of a 200-watt, 120-volt incandescent lamp?
   a. 50 ohms
   b. 72 ohms
   c. 100 ohms
   d. 333 ohms

2. What voltage is required to power a 25-watt, 5-ohm resistor?
   a. 5 volts
   b. 10 volts
   c. 11.2 volts
   d. 120 volts

3. A 75-watt lightbulb generates 30 watts of heat. What is the efficiency of the lightbulb?
   a. 60.0%
   b. 40.0%
   c. 66.75%
   d. 100.0%

4. A 120-volt circuit is drawing 5 amperes. How much power is it using?
   a. 24 watts
   b. 120 watts
   c. 350 watts
   d. 600 watts

5. A 100-watt lightbulb is 95% efficient. How much heat is dissipated by the bulb in watts?
   a. 95 watts
   b. 5 watts
   c. 25 watts
   d. 50 watts

6. What is the resistance of a 50-watt, 5-ampere circuit?
   a. 10 ohms
   b. 5 ohms
   c. 2 ohms
   d. 1 ohm

7. A 150-watt lightbulb has a resistance of 96 ohms. How much current does it draw?
   a. 5 amperes
   b. 1.25 amperes
   c. 1.56 amperes
   d. 1 ampere

8. How many foot-pounds per second is a 200-horsepower motor?
   a. 149,200
   b. 272
   c. 109,706
   d. 200

9. In one day, a certain motor uses 48 kilowatt-hours. If the energy consumption is constant all day long, how many horsepower is the motor?
   a. 2,000
   b. 2.68
   c. 83
   d. 700

10. How many BTUs per hour is a 400-horsepower motor?
    a. 937,334
    b. 298,400
    c. 746
    d. 1,050

PART

2

DC SERIES CIRCUITS

chapter 5

Understanding and Calculating Resistance in a DC Series Circuit

■ OUTLINE

■ OVERVIEW

Electrical circuits can be classified into three basic formats—series, parallel, and combination. In this chapter you will learn about the first type: series circuits. A series circuit is one in which all the various components are connected together, one after another. That is, a series circuit is one in which the same current flows through all the devices.

Consider the two resistors shown in Figure 5–1a. If the resistors are connected to a voltage source as shown in Figure 5–1b, they are said to be in series. Notice that the same current flow is present in both resistors.

If, however, the resistors are connected as shown in Figure 5–1c, they are said to be connected in parallel, and the current through each one is different, even though the total current will be equal to the sum of the two resistor currents ($I_t = I_1 + I_2$). This chapter will deal with circuits that are similar to those shown in Figure 5–1b.

Before you can analyze circuits, you must understand something about the symbols that are used for drawing them. The first part of this chapter will discuss the symbols that are used to represent various types of components and how to interpret schematic diagrams in terms of the component drawing.

■ OBJECTIVES

After completing this chapter, you should be able to:

1. Identify which specific component of a circuit is being represented by a specific symbol on a schematic drawing.
2. Draw the correct symbol for electrical or electronic components when making schematic drawings.
3. Describe the type and construction of various standard resistors. Include the resistance value, wattage rating, and tolerance in your description.
4. Determine a resistor value and tolerance by using its color bands.
5. Calculate the total resistance in series circuits using both the formulas for series resistance and Ohm's law.
6. Describe the procedures for measuring resistance using a multimeter.

FIGURE 5–1   Resistors in series and parallel: (a) resistors; (b) series; (c) parallel.

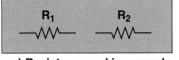

a) Resistors used in example

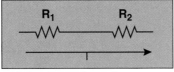

b) Resistors in series

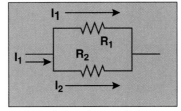

c) Resistors in parallel

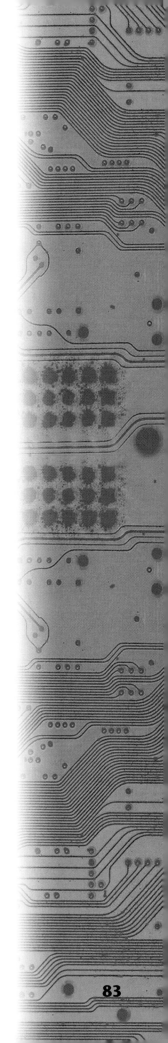

■ GLOSSARY

Component diagram   A drawing that shows the interconnection of system components by using photographs or drawings of the actual components.

Composition carbon resistor   A resistor that derives its resistance from a combination of carbon graphite and a resin bonding material.

Metal film resistor   A resistor that derives its resistance from a thin metal film applied to a ceramic rod.

Metal glaze resistor   Similar to a metal film resistor except the film is much thicker and is made of metal and glass.

Pictorial diagram   See Component diagram.

Schematic diagram   A structural or procedural diagram, especially of an electrical or mechanical system,[1] using special symbols to represent the actual physical components.

Wirewound resistor   A resistor that is made by winding resistive wire around an insulating form.

■ DIAGRAMS

The world of DC circuits can be described as a "bunch of wires and components all connected together that somehow make sense." Of course, that would not be a very accurate description of how the DC circuits actually work. If you've ever looked at any DC circuits, you've probably noticed a lot of components that look the same from circuit to circuit. There are also a large number of components that are different: Some are very different in size and rating, while others are only slightly different. It is learning these differences that will aid you in becoming a professional electrical worker.

5.1 Component Diagrams

One of the first steps is learning how the components are represented in circuit drawings. One method of drawing the circuit is to actually draw the components themselves, as shown in Figure 5–2.

This drawing is a representation of the components in a flashlight and shows four distinct components:

1. The switch that is used to turn the lamp on and off.
2. The batteries that are used to supply power to the circuit.
3. The lamp that emits visible light when current flows through it.
4. The wires or conductors that carry the current throughout the circuit.

Component diagrams are sometimes used to show the physical relationship between components or to show components of a circuit when the reader might not be familiar with those components. Component diagrams are also called **pictorial diagrams**.

Figure 5–3 is a **schematic diagram** of the same flashlight. A schematic diagram shows a circuit in such a way as to make it easy to follow the flow of current and see how the circuit is actually connected. You will notice that the schematic uses symbols instead of a drawing or picture of the actual component. The only component that is the same as in Figure 5–2 is the wire or conductors. All others are represented by their schematic symbols.

The schematic symbols represent the components in a standard way using common figures. These symbols allow you to interpret the circuit drawings. As you use schematics more and more, it will become easier for you to trace the current paths through the circuit. Tracing the main and alternate current paths is a skill you will need over and over throughout your career as an electrical worker.

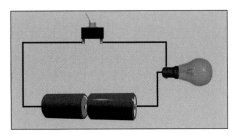

FIGURE 5–2   Circuit component diagram.

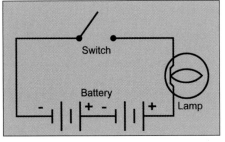

FIGURE 5–3   Schematic diagram.

■ SCHEMATIC SYMBOLS

Before you can easily interpret a schematic, you must be able to identify each component. Many components have more than one symbol to represent them, depending on the type of schematic used. The symbols you will study in this lesson will be used throughout the DC theory course. Table 5–1 provides a short description with each symbol to help you understand where that component may be used in a circuit.

Table 5–1   Schematic Symbols for Common Electrical Components

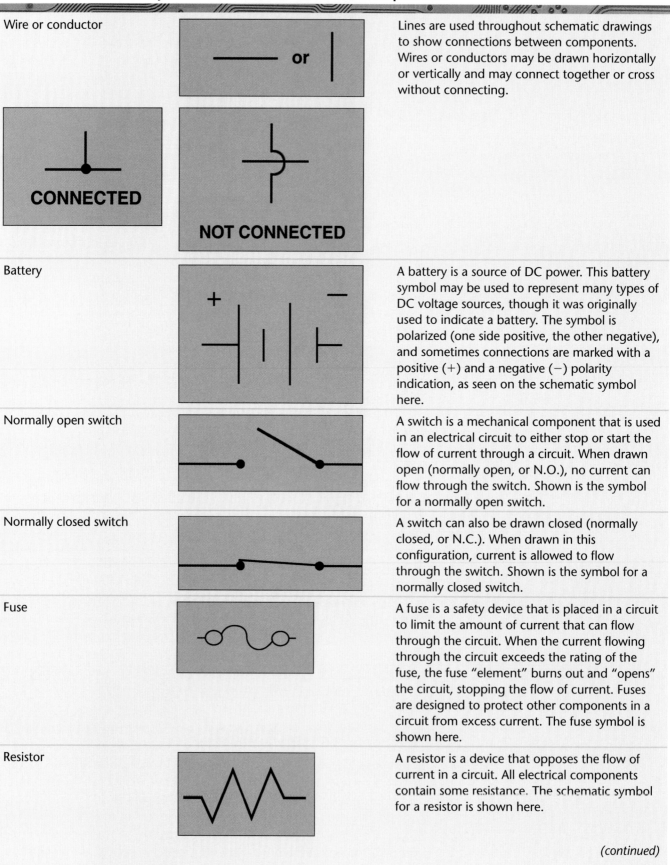

Wire or conductor	— or |	Lines are used throughout schematic drawings to show connections between components. Wires or conductors may be drawn horizontally or vertically and may connect together or cross without connecting.
	CONNECTED / **NOT CONNECTED**	
Battery	+ —	A battery is a source of DC power. This battery symbol may be used to represent many types of DC voltage sources, though it was originally used to indicate a battery. The symbol is polarized (one side positive, the other negative), and sometimes connections are marked with a positive (+) and a negative (−) polarity indication, as seen on the schematic symbol here.
Normally open switch		A switch is a mechanical component that is used in an electrical circuit to either stop or start the flow of current through a circuit. When drawn open (normally open, or N.O.), no current can flow through the switch. Shown is the symbol for a normally open switch.
Normally closed switch		A switch can also be drawn closed (normally closed, or N.C.). When drawn in this configuration, current is allowed to flow through the switch. Shown is the symbol for a normally closed switch.
Fuse		A fuse is a safety device that is placed in a circuit to limit the amount of current that can flow through the circuit. When the current flowing through the circuit exceeds the rating of the fuse, the fuse "element" burns out and "opens" the circuit, stopping the flow of current. Fuses are designed to protect other components in a circuit from excess current. The fuse symbol is shown here.
Resistor		A resistor is a device that opposes the flow of current in a circuit. All electrical components contain some resistance. The schematic symbol for a resistor is shown here.

(continued)

Voltmeter	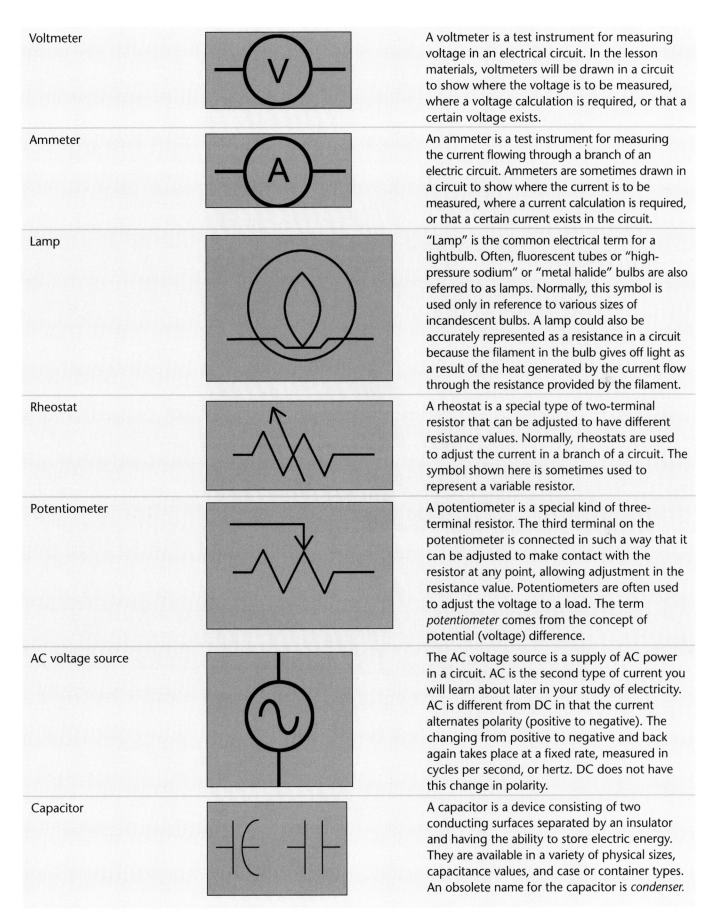	A voltmeter is a test instrument for measuring voltage in an electrical circuit. In the lesson materials, voltmeters will be drawn in a circuit to show where the voltage is to be measured, where a voltage calculation is required, or that a certain voltage exists.
Ammeter		An ammeter is a test instrument for measuring the current flowing through a branch of an electric circuit. Ammeters are sometimes drawn in a circuit to show where the current is to be measured, where a current calculation is required, or that a certain current exists in the circuit.
Lamp		"Lamp" is the common electrical term for a lightbulb. Often, fluorescent tubes or "high-pressure sodium" or "metal halide" bulbs are also referred to as lamps. Normally, this symbol is used only in reference to various sizes of incandescent bulbs. A lamp could also be accurately represented as a resistance in a circuit because the filament in the bulb gives off light as a result of the heat generated by the current flow through the resistance provided by the filament.
Rheostat		A rheostat is a special type of two-terminal resistor that can be adjusted to have different resistance values. Normally, rheostats are used to adjust the current in a branch of a circuit. The symbol shown here is sometimes used to represent a variable resistor.
Potentiometer		A potentiometer is a special kind of three-terminal resistor. The third terminal on the potentiometer is connected in such a way that it can be adjusted to make contact with the resistor at any point, allowing adjustment in the resistance value. Potentiometers are often used to adjust the voltage to a load. The term *potentiometer* comes from the concept of potential (voltage) difference.
AC voltage source		The AC voltage source is a supply of AC power in a circuit. AC is the second type of current you will learn about later in your study of electricity. AC is different from DC in that the current alternates polarity (positive to negative). The changing from positive to negative and back again takes place at a fixed rate, measured in cycles per second, or hertz. DC does not have this change in polarity.
Capacitor		A capacitor is a device consisting of two conducting surfaces separated by an insulator and having the ability to store electric energy. They are available in a variety of physical sizes, capacitance values, and case or container types. An obsolete name for the capacitor is *condenser*.

Table 5–1   *continued*

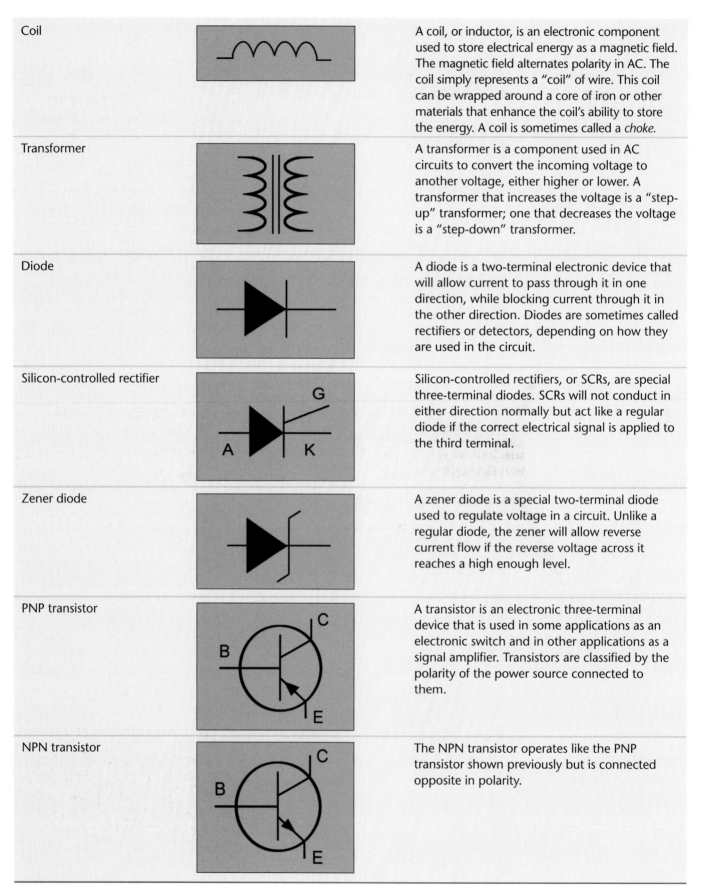

Coil	A coil, or inductor, is an electronic component used to store electrical energy as a magnetic field. The magnetic field alternates polarity in AC. The coil simply represents a "coil" of wire. This coil can be wrapped around a core of iron or other materials that enhance the coil's ability to store the energy. A coil is sometimes called a *choke*.
Transformer	A transformer is a component used in AC circuits to convert the incoming voltage to another voltage, either higher or lower. A transformer that increases the voltage is a "step-up" transformer; one that decreases the voltage is a "step-down" transformer.
Diode	A diode is a two-terminal electronic device that will allow current to pass through it in one direction, while blocking current through it in the other direction. Diodes are sometimes called rectifiers or detectors, depending on how they are used in the circuit.
Silicon-controlled rectifier	Silicon-controlled rectifiers, or SCRs, are special three-terminal diodes. SCRs will not conduct in either direction normally but act like a regular diode if the correct electrical signal is applied to the third terminal.
Zener diode	A zener diode is a special two-terminal diode used to regulate voltage in a circuit. Unlike a regular diode, the zener will allow reverse current flow if the reverse voltage across it reaches a high enough level.
PNP transistor	A transistor is an electronic three-terminal device that is used in some applications as an electronic switch and in other applications as a signal amplifier. Transistors are classified by the polarity of the power source connected to them.
NPN transistor	The NPN transistor operates like the PNP transistor shown previously but is connected opposite in polarity.

Table 5–1   *continued*

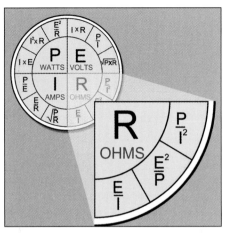

FIGURE 5–4 Ohm's law pie chart.

■ THE SERIES CIRCUIT

5.2 Ohm's Law

In chapter 3, you learned about Ohm's law and used a pie chart (see Figure 5–4) to discuss the different relationships among resistance, current, voltage, and power. In this chapter, you will concentrate on the resistance slice of the pie chart. Applications of these resistive relationships are presented in a series circuit.

A series circuit is one in which there is only one path for current to flow. In Figure 5–5, all four resistors are connected end to end so that all the electrons leaving one resistor are entering the next. This flow continues through all resistors, the switch, fuse, and battery. Even though each resistor in a series circuit offers some opposition to the flow of current in the series circuit, the current flow is the same throughout the circuit. If you were to take an ammeter (a meter designed to measure current flow) and place it in series with the components of the series circuit by opening the circuit and connecting the leads of the ammeter, the amount of current measured would be the same regardless of where you opened the circuit to connect the meter. This is a very important concept to remember. **The current is the same throughout the series circuit.**

Each resistor in the series circuit adds resistance and impedes the current flow throughout the circuit. The connecting wires, the fuse, and the battery provide some resistance, but compared to the resistors in the circuit, they account for a very small amount of the total resistance and are neglected in most calculations.

5.3 Calculating Resistance in a Series Circuit

All components in a circuit provide some amount of resistance to the current flow. The resistors, shown in Figure 5–5, provide almost all the total circuit resistance. The connecting wires, the fuse, and the battery are designed to give as little resistance as possible. To keep the concepts simple, we will assume that the total resistance in Figure 5–5 is due only to the resistors. As such, the total circuit resistance is the sum of the individual resistors and can be shown in Equation 1:

$$R_T = R_1 + R_2 + R_3 + R_4 + \dots R_n \qquad (1)$$

where n represents the total number of resistors in the circuit (in Figure 5–5, $n = 4$) and R_T = total resistance.

FIGURE 5–5 Simple series circuit.

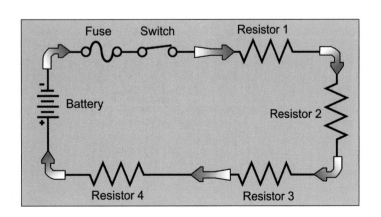

FIGURE 5–6   Simple series circuit with component values shown.

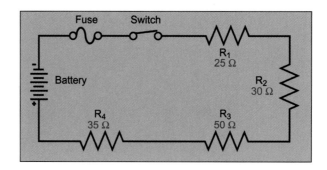

Now apply values to the circuit as in Figure 5–6.

EXAMPLE 1

What is the total resistance of the series DC circuit in Figure 5–6?

Solution:
Using Equation 1;

$$R_T = R_1 + R_2 + R_3 + R_4$$
$$R_T = 25 \ \Omega + 30 \ \Omega + 50 \ \Omega + 35 \ \Omega$$
$$R_T = 140 \ \Omega$$

EXAMPLE 2

Calculate the total resistance of the circuit in Figure 5–7.

Solution:
Using Equation 1;

$$R_T = R_1 + R_2 + R_3$$
$$R_T = 5 \ \text{k}\Omega + 150 \ \Omega + 3.5 \ \text{k}\Omega$$
$$R_T = 5,000 \ \Omega + 150 \ \Omega + 3,500 \ \Omega$$
$$R_T = 8,650 \ \Omega$$

or

$$R_T = 8.65 \ \text{k}\Omega$$

Another way to calculate the total resistance is to use Ohm's law (refer to Figure 5–4). To do this, you must know two of the following:

FIGURE 5–7   Another example of a series circuit.

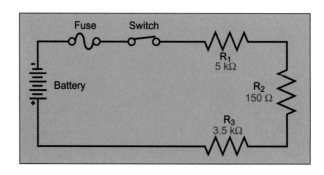

1. The total current
2. The total voltage drop across all the resistors
3. The total power

When you have two of these three, it is simply an application of the equations used in Ohm's law pie chart.

EXAMPLE 3

Calculate the total resistance of the circuit in Figure 5–8 using Ohm's law.

Solution:
Using Ohm's law equation for resistance;

$$R_T = \frac{E}{I}$$

$$R_T = \frac{30 \ V}{0.25 \ A}$$

$$R_T = 120 \ \Omega$$

EXAMPLE 4

Calculate the total resistance for the circuit in Figure 5–9 using Ohm's law.

Solution:
Using Ohm's law and the equation for power, we get the following:

$$R_T = \frac{E^2}{P}$$

$$R_T = \frac{28^2}{7}$$

$$R_T = \frac{784}{7}$$

$$R_T = 112 \ \Omega$$

In a later chapter you will learn how to calculate the individual values of resistance for R_1 and R_2 in Figure 5–9.

FIGURE 5–8   Calculating total resistance in a series circuit using current and voltage.

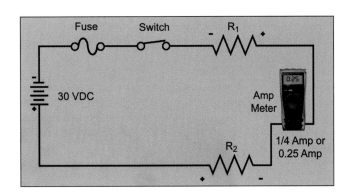

FIGURE 5-9 Calculating total resistance in a series circuit using voltage and power.

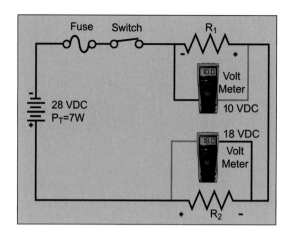

■ TYPES OF RESISTORS

5.4 Overview

There are two general types of resistors: fixed and variable. Their names describe the main difference between the two. A fixed resistor has a value that is set and does not change. A variable resistor can have its value changed. In addition to the two types of resistors, there are three important numbers that are associated with any type of resistor: the resistance value, the power rating (wattage rating), and the tolerance.

Resistance Value

The resistance value is the measurement of the amount of opposition that the resistor offers to current flow. This value is measured in ohms and is represented by the symbol Ω (omega).

Wattage Value

The second number that is important when working with resistors is the wattage value or power rating of the resistor. The power rating of the resistor is the amount of power that can be dissipated by the resistor without damaging the resistor or affecting its operation. The power rating is measured in watts. Using a resistor with too low a power rating can result in the destruction of the device or variations in the resistor value due to the effects of heat generated by the device itself.

Tolerance

Variations in resistance values may show up among resistors. Even resistors of exactly the same type and specified resistance will be somewhat different due to variations in material, quality control, and other such issues. This means that if you measure the resistance of 100 resistors of the same type and specified resistance, you will find that they are all slightly different from one another. The allowed maximum variation is called the tolerance and is equal to the allowed percentage variation in the resistor's specified resistance.

For example, if a family of resistors are specified as $100\,\Omega \pm 2\%$. The measured value of any one of those resistors could be as low as 98 Ω or as high as 102 Ω. Any value in this range for that type of resistor is acceptable. Common resistance tolerances include 1%, 2%, 5%, 10%, and 20%. Higher precisions are available if needed.

A resistor that has been manufactured for a 2% tolerance and measures 104 Ω would have been acceptable under the 20% tolerance requirement. However, resistor tolerances are marked on many resistor cases by ring color, numerical value, or some other identifiable code. As today's sophisticated electrical and electronic equipment continues to evolve, it is important to initially choose and maintain consistency in the replacement of components of the exact same design tolerance for which they were manufactured.

5.5 Fixed Resistors

Construction

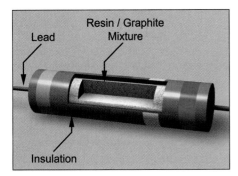

FIGURE 5–10 Composition carbon resistor (fixed).

One of the most common types of fixed resistors is the **composition carbon resistor**. It is made of a compound of carbon graphite and a bonding resin. The value of the resistor is determined by the ratio of the resin to the carbon graphite. An insulating material surrounds the mixture (see Figure 5–10). They are the most popular resistors because they are durable, cheap, and readily available. However, they can change value with age or overheating. They are made in a wide range of values and ratings. The wattage (power) rating is determined by their size: 0.5-W resistors are ⅛ of an inch in diameter and ⅜ of an inch long, 1-W resistors are ¼ of an inch in diameter and ⁷⁄₁₆ of an inch long, and 2-W resistors are ⁵⁄₁₆ of an inch in diameter and ¹¹⁄₁₆ of an inch long.

FIGURE 5–11 Metal film resistor (fixed).

Another type of fixed resistor is the **metal film resistor** (see Figure 5–11). These are made by applying a metal film to a ceramic rod in a vacuum. The resistance value is determined by the thickness of the metal film applied and the type of metal used. The thickness of the metal film can range from 0.00001 to 0.00000001 of an inch. Metal film resistors are better than the composition carbon resistor because their values change less over time and they have a closer tolerance. While composition carbon resistors can deviate as much as 20%, metal film resistors vary from 0.1% to 2%. The manufacturing costs of metal film resistors are much higher than composition carbon types.

The **metal glaze resistor** is similar to the metal film resistor. It is constructed by combining metal with glass and then applying this to a ceramic core as a thick film. The ratio of the metal to the glass determines the rating value of the resistor. The metal glaze resistor also has good tolerances of 1% to 2%.

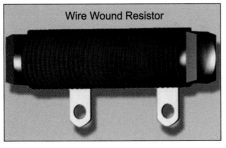

FIGURE 5–12 Wirewound resistor (fixed).

Wirewound resistors (see Figure 5–12) are a resistive wire wound around an open core and may or may not be covered with a protective outer layer or insulating material. The wirewound resistors are best used in applications requiring high temperature and high power. They can withstand more heat than any other type of resistor. They are expensive to construct and require a large amount of space for mounting. In addition, because it is a coil, a wirewound resistor will add a certain amount of inductance in higher-frequency circuits. Inductance will be covered in the AC portion of your training.

FIGURE 5–13 Standard color coding chart for resistors.

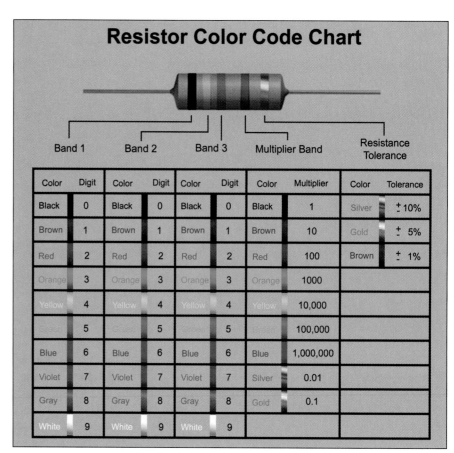

Resistor Color Code Chart

Color	Digit	Color	Digit	Color	Digit	Color	Multiplier	Color	Tolerance
				Band 1		Band 2		Band 3	Multiplier Band	Resistance Tolerance

Color	Digit	Color	Digit	Color	Digit	Color	Multiplier	Color	Tolerance
Black	0	Black	0	Black	0	Black	1	Silver	± 10%
Brown	1	Brown	1	Brown	1	Brown	10	Gold	± 5%
Red	2	Red	2	Red	2	Red	100	Brown	± 1%
Orange	3	Orange	3	Orange	3	Orange	1000		
Yellow	4	Yellow	4	Yellow	4	Yellow	10,000		
Green	5	Green	5	Green	5	Green	100,000		
Blue	6	Blue	6	Blue	6	Blue	1,000,000		
Violet	7	Violet	7	Violet	7	Silver	0.01		
Gray	8	Gray	8	Gray	8	Gold	0.1		
White	9	White	9	White	9				

FIGURE 5–14 Variable resistor.

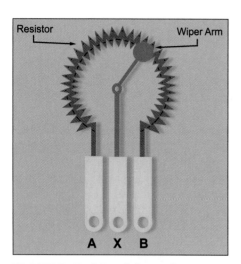

FIGURE 5–15 Schematic diagram of a variable resistor.

CAUTION: Wirewound resistors, when operated under normal conditions, can run hot enough to inflict serious injuries or burns from the heat generated by the device. One of the uses of the wirewound resistor is as a heating element whose sole purpose is to generate heat to surrounding equipment or areas.

Color Coding

Fixed resistors have a color code to indicate the value of the resistor using the colored rings around the resistor. Figure 5–13 describes the coding. By looking at the bands on the resistor and using the chart in Figure 5–13, the value of the resistor can be determined. For example, if the colors on the resistor were in the order black, yellow, orange, green, and gold, the resistor digits would be 0 = black, 4 = yellow, 3 = orange, a multiplier of 100,000 (or five zeros) = green, and a tolerance of 5%. This resistor would have a value of 4,300,000 ohms, or 4.3 megohms.

5.6 Variable Resistors

Construction

Variable resistors are resistors whose resistance value can be changed by some method. Figure 5–14 shows a standard variable resistor. Figure 5–15 shows a schematic of its electrical components. A stem connected to a wiper arm controls the variance in resistance values. As the wiper arm is turned, the resistance measured between

FIGURE 5-16 Multiturn variable resistor.

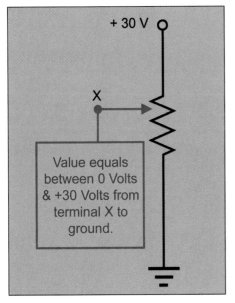

FIGURE 5-17 Schematic of a potentiometer.

FIGURE 5-18 Schematics of various types of resistors.

the wiper arm terminal and the terminal at the end of the resistor changes. Such resistors are sometimes called *potentiometers.*

Variable resistors may be either composition or wirewound. Like fixed wirewound resistors, variable wirewound resistors can handle more power than composition resistors. Such wirewound resistors could be used to control large currents in devices such as battery chargers or electroplating processes.

The resistance between points A and B in the variable resistor pictured in Figure 5–15 is fixed. The value of the resistance between points A and X (AX) or B and X (BX) varies, depending on where the wiper arm is positioned. The picture in Figure 5–15 would indicate that points AX have a higher resistance than points BX. This is because there is more resistive material between the contacts of AX than between the contacts of BX.

Figure 5–16 shows an example of a multiturn variable resistor. These are operated using a knob or screw. For example, a five-turn multiturn resistor would require five full rotations to go from the least to the highest resistance value. These types of resistors provide much finer control and are used in precision applications, such as the alignment of electronic devices.

Potentiometers and Rheostats

Some of the more common terms you will encounter in your career as an electrician are *potentiometer* and *rheostat.* A potentiometer is a variable resistor that has three terminals. A rheostat is a variable resistor that has two terminals. A potentiometer may be used as a rheostat if only two of the three terminals are used. Figure 5–17 shows a schematic representation of a variable resistor being used as a potentiometer.

Potentiometers are generally used to adjust the voltage applied to a circuit, while a rheostat is used to adjust the level of current through a circuit.

Tapped Resistors

A tapped resistor is a fixed resistor with a tap (or taps) permanently connected between the terminals of a fixed resistor. The resistance value between each tap or terminal is fixed. Therefore, depending on which tap you use, you can vary the resistance value. However, the resistance between each possible connection point remains fixed.

Resistor Symbols

Figure 5–18 shows some of the common schematic representations for popular types of resistors.

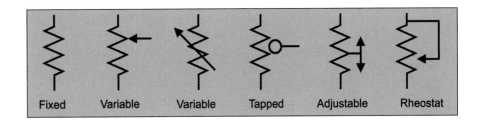

■ USING AN OHMMETER

A meter designed to measure resistance is called an *ohmmeter*. Meters designed to measure several circuit characteristics are called *multimeters*. A multimeter generally contains an ohm scale for reading resistance. Resistance is a physical characteristic of a conductor or anything conductive and cannot be measured directly. Ohmmeters and multimeters that measure resistance do so by applying a known voltage across the resistor and measuring the resultant current or by supplying a known current to a resistor and then measuring the resultant voltage. These are both applications of Ohm's law. Scales on ohmmeters are calibrated to read resistance rather than voltage or current. Multimeters have multiple scales, each having a maximum resistance value, calibrated in ohms, that is used for resistance measurements.

Resistances in circuit components vary from very small to very large values. For different applications, it is sometimes necessary to measure resistance values that vary from micro-ohms (.000001) to megohms (1,000,000). It is not practical to design a meter that could read such a large range of resistance values on a single scale. Most ohmmeters available today have a range selector switch that allows the user to select the appropriate range for reading the resistance values being measured.

5.7 Analog Meters

Figure 5–19 is an analog multimeter. The analog multimeter uses a moving indicator and a fixed set of graduated scales. As the value being measured changes, the meter needle or pointer will move to some location between the meter scale limits. The value being measured can then be read from the appropriate scale being used. Multimeters are capable of measuring voltage, current, and resistance. The top scale on the meter in Figure 5–19 is the ohms scale used for resistance measurements. A selector switch allows the user to select three different ohms scales ranges. All ohm ranges use the same (top) scale. The need

FIGURE 5–19   Analog multimeter.

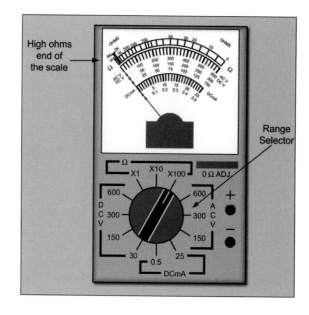

FIGURE 5–20 Resistance scales for the analog meter shown in Figure 5–19.

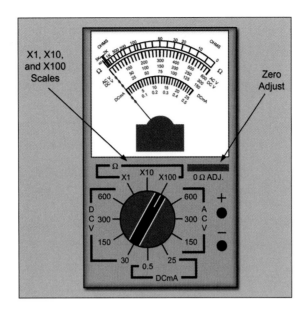

for multiple resistance scales is necessary because of the nonlinearity of the ohm scale. Even though the scale goes from 0 ohms to ∞ (infinity, or an open circuit) ohms, the scale is very difficult to read accurately at the upper end of the scale. Notice how the numbers are closer together at the left, or high end, of the scale.

The analog meter includes a selector switch that can be used to select scales ×1, ×10, and ×100 (see Figure 5–20). This makes reading the correct value much easier since a scale can be selected to match the value of the resistance to be measured. If the pointer reads close to either end of the scale, selecting a different scale may allow the user to move the needle toward the center of the meter's scale. This provides for increased accuracy in reading the resistance value.

When the selector switch on the face of the meter is in the ×1 position, the meter's scale is read directly, and simply reading the value indicates the correct resistance.

When the selector is in the ×10 position, multiply the value on the scale faceplate indicated by the pointer by a factor of ten. In the ×10 position, for instance, if the pointer indicated a resistance of 30, the value would actually represent a resistance of 10 × 30, or 300 ohms. Similarly, when the selector switch is in the ×100 position, the value on the faceplate is multiplied by 100. In the ×100 position and the pointer indicating 2 k (or 2,000 ohms), the actual reading would be 2 k × 100, or 200 kΩ.

A "0 ΩADJ." control knob allows the meter to be adjusted to compensate for test lead resistance and battery voltage level when making resistance measurements. To adjust this control, the user would "short" the leads by touching them together and then adjust the control to give a "0 Ω" reading on the faceplate. This adjustment eliminates the resistance of the test leads from any resistance reading taken with the meter and must be done every time the meter is used and whenever the scale is changed from one range to another. The zero on the meter should be checked periodically, as the zero value may drift as the batteries weaken.

FIGURE 5–21 Digital multimeter.

5.8 Digital Meters

Figure 5–21 shows a digital multimeter that can be used to measure resistance. Digital meters are read differently than analog meters. As you can see, the moving pointer and multiscale faceplate are no longer present. In the digital multimeter, they have been replaced by a liquid crystal display (LCD).

When using a digital meter, a single value is digitally displayed that represents the resistance being measured. A selector switch is used to select the maximum resistance to be read on the selected scale. The amount of information displayed depends on the type of display shown on the meter. Meters may display two and one-half digits (199), three digits (999), three and one-half digits (1999), or four digits (9999).

The meter shown is classified as a three-and-one-half-digit meter. This means the display has three digits that can show any value for 0 to 9 and a fourth digit that can indicate either a 0 or a 1. The maximum reading for this meter is 1999. A decimal point is used to scale this reading. Scales shown for this meter include 2 K, 200, 20 K, 200 K, 2 M, and 20 M. These scales represent the maximum resistance readable for that scale and would be displayed as shown in Table 5–2.

On the digital meter, when a value exceeds the maximum value that can be displayed for that scale, the meter indicates the overscale condition by displaying a 1 or a 9 or by flashing 9999. When the value of resistance being measured represents a resistance value below the scale of the meter (such as might occur if a 200-Ω resistor was to be checked while the meter was set on the 2-M scale), then the meter will indicate the underscale condition by displaying a 0.00.

When measuring resistance on the digital meter shown in Figure 5–18, if you encounter a 1, you should increase the scale reading (turn the selector knob counterclockwise on the meter shown) to bring the meter within scale. If you encounter a 0.00 on the display, you should decrease the scale reading (turn the selector knob clockwise) until you have the correct reading. Other meters may differ. Be sure to understand how the meter you are using operates.

One additional resistance scale, indicated by the symbol)))), is used for continuity tests. A continuity test is used to indicate a current path, "continuity," exists between two points. This type of test is used

Table 5–2 Digital Multimeter Scales

Scale	Maximum Reading	Display at Maximum
200	200 Ω	199.9
2 K	2,000 Ω	1.999
20 K	20,000 Ω	19.99
200 K	2,000,000 Ω	199.9
2 M	2,000,000 Ω	1.999
20 M	20,000,000 Ω	19.99

when the user is concerned not about the specific resistance in the current path but only that a continuous current path exists in the circuit. When set to this scale, the meter will indicate a 1 for an open circuit and a 000 for a closed or shorted circuit. In addition, with some meters, an audible signal is heard when a closed circuit is encountered.

The digital multimeter does not have a "0 Ω ADJ." control, as zero adjustment is normally not required for digital multimeters. Some digital multimeters have autoranging capabilities that automatically adjust the meter's range to suit the value being measured.

5.9 Measuring Resistance

When measuring resistance with either type of meter, the procedure to obtain accurate measurement of the value of resistance is the same. The digital meter is easier to read when the resistance of a specific component is required. Digital meters are less desirable, however, when reading the resistance of circuit components that have a quickly changing resistance (such as a resistive sensor in which the resistance depends on some physical variable, such as light level or temperature). The main reason for this is that the variable being measured is rapidly changing and the process required within the digital meter to read, interpret, and display the measured value often exceeds the time between the changes in that variable. This does not mean that the digital meter is slow; rather, it may be inappropriate for the type of variable being measured. For these types of quickly changing values, the analog meter is often a better choice. Note that many modern digital meters are equipped with an LCD analog bar that can be used for tracking changes.

Since meters that measure resistance have their own power sources, all power must be removed from circuits before attempting to measure resistance. Circuit power sources not only will affect the reading on resistance measuring devices but also might pose a safety hazard if left connected.

CAUTION: Never attempt to measure the resistance of a component in a circuit while power is applied to that circuit. To do so will not only damage or destroy the meter used for the test but also may cause harm or personal injury to the individual performing the test.

Other components connected in a circuit might also affect the accuracy of resistance measurements. When measuring resistance, you must verify that the only current path available is the one through the component whose resistance you wish to measure.

Generally, when measuring the resistance of components in a circuit, the polarity of the ohmmeter is not important. Most resistive components are not sensitive to the polarity of the voltage applied to those components during a resistance test. Electronics devices such as diodes and transistors are polarized; therefore, caution must be applied to ensure that the polarity of the meter's internal source is correct for the test to be accurate.

Figure 5–22 shows the proper way to measure the resistance of resistor R_1. Both the power source and all other current paths have been isolated from the component whose resistance is to be measured.

FIGURE 5-22 Proper method for resistance measurement.

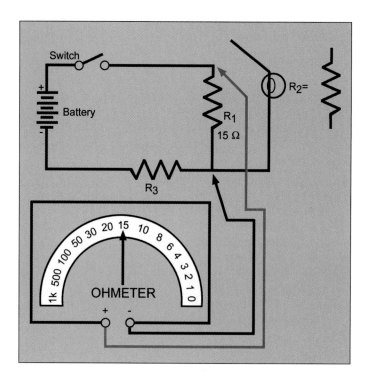

FIGURE 5-23 Improper (and possibly dangerous) method for resistance measurement.

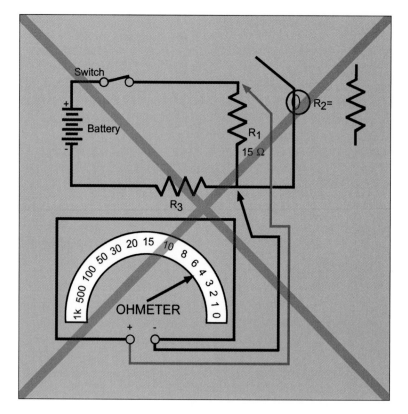

Figure 5–23 shows a circuit in which the resistance measurement would be in error because the power is still connected to the circuit. The reading would show that the resistance of R_1 was higher (or lower) than the actual value because of the presence of the circuit power source, which would subtract from (or add to) the ohmmeter's own internal power source.

FIGURE 5–24   Improper method for resistance measurement.

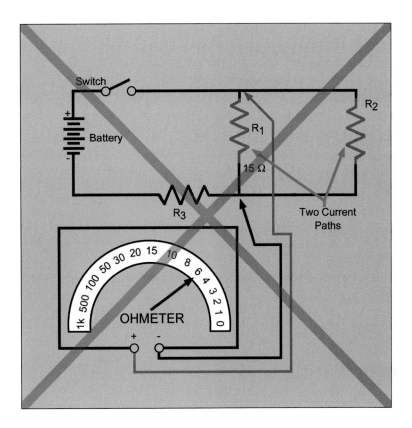

CAUTION: This connection could also be dangerous. Do not do this.

Figure 5–24 shows a circuit in which the resistance measurement would be in error because more than one current path exists for current to flow in the circuit. The internal current source from the ohmmeter would flow through both R_1 and R_2, and the resistance reading would represent the equivalent resistance of resistors R_1 and R_2 combined, which would be lower than the resistance of R_1 alone.

■ SUMMARY

The analysis of electrical circuits is greatly aided by the use of appropriate diagrams. One of the often used and most convenient electrical diagrams is the schematic diagram. The schematic diagram uses standardized symbols to show the interconnection of the various components in such a way that the voltages, currents, resistances, and powers can be calculated by careful circuit analysis. The electrician must know what each of the symbols means so that he or she can properly draw, read, and analyze electrical circuits.

One of the most commonly used components in an electrical circuit is the resistor. The resistor is a component that opposes the current flow in the circuit. Resistors are made in a variety of different ways,

including fixed (composition, carbon, metal film, metal glaze, and wirewound) and variable. Each has its own advantages and disadvantages and is used accordingly. Whatever the type, resistors are designed to exhibit specific amounts of resistance for specific purposes.

Total resistance in a series circuit can be calculated in a number of different ways. Perhaps the most obvious is to simply add the value of each individual resistor to get the total. If the individual values are not known, the total resistance can be determined if any two of the voltage, current, and power are known.

It is also possible to determine the resistance of individual resistors if the resistor color bands are visible. Both the resistance and the tolerance of the resistor can be determined. Once the individual values are known in a series circuit, the total can be derived by adding the individual resistors as before.

If no color codes are available and no circuit values can be determined (volts, amperes, or power), the electrician can still determine circuit resistance values by using an ohmmeter. The ohmmeters that are most often used in modern electrical work are those that are part of an analog or digital multimeter. In this chapter you learned that such measurements can be made, but they must be made carefully to avoid erroneous readings and/or safety hazards.

■ REVIEW QUESTIONS

1. Review the circuit symbols shown in Table 5–1. Draw a circuit that shows a battery in series with an ammeter, in series with three resistors, and in series with a fuse.

2. Describe the differences between a series circuit of two resistors and a parallel circuit of the same two resistors.

3. Compare and contrast carbon-composition resistors, metal film resistors, and wirewound resistors. The elements to include in your discussion are resistor values, tolerances, and power capabilities.

4. Carefully review Figure 5–13. What would be the color code for a 2,200-ohm resistor with ±1% tolerance? A 5-megohm resistor with ±20% tolerance?

5. Discuss the various uses for variable resistors. What advantage does a ten-turn variable resistor have over a one-turn variable resistor?

6. How can a variable resistor be used to vary the voltage applied to some circuit part? (potentiometer)

7. You are using the meter shown in Figure 5–21. You make a resistor measurement, and the display shows a 1. What is wrong, and how can you fix it?

8. Explain why Figure 5–24 is an improper way to measure the resistance of resistor R_1.

9. Explain why the measurement shown in Figure 5–23 might be dangerous.

10. Look at Figure 5–6. All the elements in this circuit exhibit some resistance. Why can you generally ignore the resistance of the wire, battery, fuse, and switch?

■ PRACTICE PROBLEMS

1. In Figure 5–25, what is the resistance and tolerance of each resistor?

2. In Figure 5–9, assume that the battery voltage is 125 volts and the total power dissipation is 25 watts. What is the total resistance of the circuit?
   a. 5 ohms
   b. 20 ohms
   c. 550 ohms
   d. 625 ohms

3. In Figure 5–9, if the battery voltage is 125 volts and the total current is 10 amperes, what is the total resistance in ohms?
   a. 1.25 ohms
   b. 12.5 ohms
   c. 125 ohms
   d. 10 ohms

4. In Figure 5–6, what is the total resistance for resistors with the value shown in the following table? All resistances are given in ohms.

R_1	R_2	R_3	R_4	R_T
5	10	15	20	
2,200	3,300	14,000	600	
1.5	2	6	14	

5. Using Ohm's law, calculate the resistance of resistor R_1 in Figure 5–9.
   a. 0.7 ohms
   b. 1.4 ohms
   c. 2.5 ohms
   d. 2.8 ohms

FIGURE 5–25 Resistor color code question.

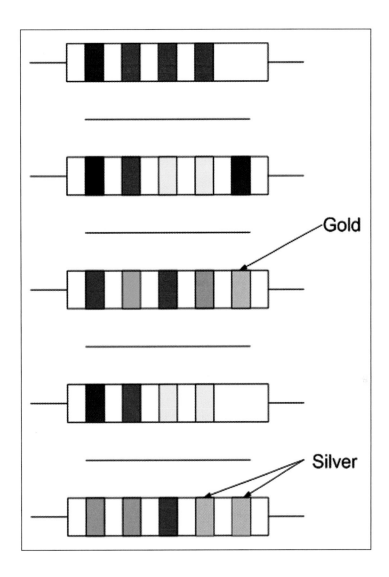

6. In Figure 5–8, if the voltage across resistor R_2 is measured as 10 volts, what is the resistance of R_2?

    a. 20 ohms

    b. 40 ohms

    c. 60 ohms

    d. 80 ohms

7. A resistor is 33,000 ohms and has a ±10% tolerance. What is the lowest resistance that you should read if you measure the resistor with a very accurate ohmmeter?

    a. 29,700 ohms

    b. 33,000 ohms

    c. 36,300 ohms

    d. 25,000 ohms

8. The voltage in a circuit is 100 volts and the current is 20 amperes. What is the resistance of the circuit?

    a. 2,000 ohms

    b. 120 ohms

    c. 5 ohms

    d. 1 ohm

9. A certain composition carbon resistor is ⁵⁄₁₆ of an inch in diameter and ¹¹⁄₁₆ of an inch long. What is its wattage rating?

    a. ¼

    b. ½

    c. 1

    d. 2

10. Look at Figure 5–9. If the top voltmeter reads 14 volts, answer the following questions:

    a. What will the bottom meter read? ____

    b. What is the value of R_1? ____

    c. R_1 is a 5-watt resistor. Will it overheat? ____

chapter **6**

How Current Reacts in a DC Circuit

■ OUTLINE

■ OVERVIEW

In chapter 5, you learned about the resistance of series circuits in DC applications. In this chapter you will learn about the way that current behaves in DC circuits and how to measure the current. You will also be introduced to some basic concepts of how to protect a circuit in the event that the current rises too high because of a short circuit or an overload.

Remember that electricity transfers energy from one point to another using current. Clearly, an understanding of how current behaves is critical in the day-to-day duties of an electrician.

■ OBJECTIVES

After completing this chapter, you should be able to:

1. Use the symbols for the fuse and circuit breaker correctly in a schematic diagram.
2. Calculate the effect of changing voltage and resistance on series circuit current.
3. Determine how some circuits may be modified to control circuit current using Ohm's law.
4. Measure the current in a series circuit using either an analog or a digital meter.

■ GLOSSARY

Circuit breaker A type of switch that can be used to open or close the circuit. It is often supplied with automatic sensing elements that will open the breaker when the current exceeds a predetermined value.

Fuse A type of protective device that features a metal link that melts when the current flow exceeds a predetermined value.

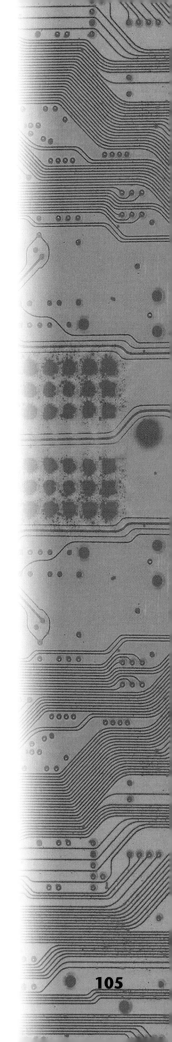

■ INTRODUCTION

In this chapter you will learn how current reacts and is affected in a DC series circuit. The current in a series circuit is the same no matter where it is measured. In Figure 6–1, the value of current passing through resistor 1 is also passing through the switch, fuse, and battery. If the current flow through resistor 1 or 2 were measured, it would be the same as through resistors 3 and 4. You can imagine the current flow as being a continuous chain of electrons, and the amount of electrons flowing past any single point in the chain is the same, no matter where you look.

Since all the current flows through every component in the circuit, each component must be considered as a load (resistive element) in determining the total current. Ohm's law defines this relationship between voltage, resistance, and current as $I_T = E/R$ (I_T = total circuit current). In the last lesson, you studied how the resistance of individual components in a series circuit add together to determine total resistance. In this lesson, you will look at how voltage and total circuit resistance are equally important in determining the total series circuit current.

■ EVALUATING AND CALCULATING DC CURRENT

6.1 Series DC Circuit Current

The most important thing to remember when determining series circuit current is that there is only one path for current flow. Note that since there is only one path, all the ammeters read the same (Figure 6–2). In this example, $I_T = I_1 = I_2 = I_3$.

This is a "current law" about a series DC circuit. All current measurements throughout the circuit will be the same. Notice also the subscripts used with the letter designations in the circuit. The subscripts allow circuit designers to accurately label each circuit component with its own unique alphanumeric label in a circuit. This is very important in more complex circuits.

There are two components in Figure 6–2 that we have not yet discussed. They are the **circuit breaker** (CB_1) and the **fuse** (F_1). Both of

FIGURE 6–1 A simple series circuit.

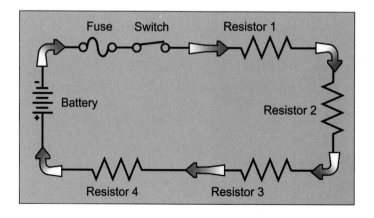

FIGURE 6–2 Current measurement in a simple series circuit.

FIGURE 6–2 Current measurement in a simple series circuit.

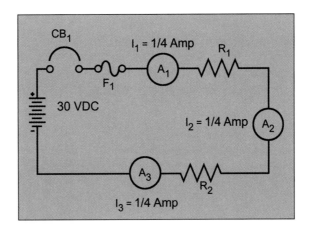

these components perform the same basic function: circuit safety. Their job is to open automatically if the amount of current in the circuit rises to an unacceptable level because of a short circuit or an overload condition. You will notice that both of these components are connected in series with all the other components. This means that each of these safety devices has the same current flowing through them as the rest of the circuit. A circuit will have a fuse, a circuit breaker, or occasionally both.

CB_1 functions to disconnect the rest of the circuit from the power source by automatically opening at a set value of current (see Figure 6–3). Most circuit breakers have to be manually reset so that the cause of the short circuit will not occur again until someone has corrected the problem. F_1 is designed to fail (melt), producing an open circuit at a preset value of current (see Figure 6–3). This prevents excessive current from damaging any other component in the circuit and, like the circuit breaker, removes the power source from the rest of the circuit. Fuses use a "sacrificial" element to protect the circuit and are not resettable. A blown (melted) fuse (or fuse element) must be replaced.

6.2 Determining Current in a Series DC Circuit

You also learned in chapter 5 that the total resistance in a series circuit is determined by the addition of all resistive components in that circuit. To keep things simple, we defined those components as just the

FIGURE 6–3 Circuit safety devices.

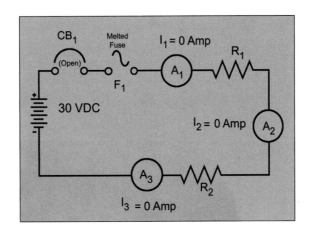

resistors. In keeping with this definition, then, resistors cause opposition to current flow, and the total opposition to current flow is the total resistance in the circuit. So, to find the total current in a circuit, we must know two things: the supply voltage (E_T) and the total resistance (R_T). This means that the total current is affected when the total resistance or the supply voltage changes. If any component increases by 10 Ω, the total resistance is increased by 10 Ω, which affects the total current.

Let's look at this using Ohm's law. The total current in a circuit is directly proportional to the total voltage (E_T) and inversely proportional to the total resistance (R_T). The formula is

$$I_T = \frac{E_T}{R_T}$$

If E_T increases, I_T increases by the same factor or rate. For example, if voltage doubles and resistance stays the same, current will double. If R_T increases, I_T decreases by the same factor: An increase in resistance by a factor of three will cause the current to decrease to one-third its original value.

Let's look at some examples using Figure 6–4a and b.

FIGURE 6–4   Two ways to calculate DC current: (a) with total voltage known and (b) with voltage across R_2 known.

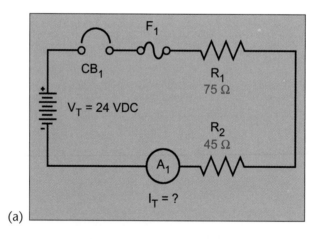

(a)

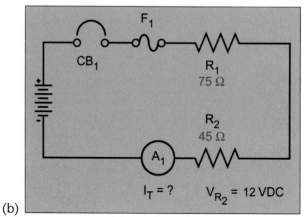

(b)

EXAMPLE 1

Calculate the total current in the circuit.

Find total resistance:

$$R_T = R_1 + R_2$$
$$R_T = 75 + 45$$
$$R_T = 120 \ \Omega$$

Using Ohm's law,

$$I_T = \frac{E_T}{R_T}$$
$$I_T = 24 \text{ V} \div 120 \ \Omega$$
$$I_T = .2 \text{ A}$$

EXAMPLE 2

Determine the value of the current (I_T) if R_2 doubles in resistance.

In this example, we have to recalculate R_T to find the new value of I_T:

$$R_T = 75 \ \Omega + (45 + 45) \ \Omega \ (R_2 \text{ doubles in value})$$
$$R_T = 75 \ \Omega + 90 \ \Omega$$
$$R_T = 165 \ \Omega$$
$$I_T = E_T \div R_T$$
$$I_T = 24 \text{ V} \div 165 \ \Omega$$
$$I_T = 0.145 \text{ A}$$

EXAMPLE 3

Apply what you have learned about series circuit current and determine the total current in the circuit Figure 6–4b.

Since the current through a series circuit is the same no matter where it is measured, the current through R_2 will be equal to the total current. So,

$$I_T = I_{R_2} = \frac{V_{R_2}}{R_2}$$
$$I_T = I_{R_2} = \frac{12 \text{ V}}{45 \ \Omega}$$
$$I_T = I_{R_2} = .266 \text{ A}$$

Remember, total resistance and the total voltage affects I_T. If either changes, recalculate the changed value and reapply Ohm's law for current.

■ USING AN AMMETER

6.3 Meter Types and Connections

In chapter 5, you learned how to use the ohmmeter scales of a multi-meter to measure resistance. When a meter is used to measure current, it is called an *ammeter*. Both the analog and the digital multimeters discussed in the lesson on resistance in a series circuit contain scales for measuring current in a circuit.

In order to measure current, the meter is inserted into and becomes part of the circuit (see Figure 6–5). While each type of meter has its own internal power source, the operation of the meter is such that the actual current supplied by the circuit is used to drive the sensing element in the meter when the meter is used to measure current. This is different than what was described in the earlier lesson on measuring resistance, where the meter supplied an internal voltage to drive the meter element.

Because of the large variations available in both voltages and resistances in a circuit, current can vary over a large range of values from a few microamps (one millionth of an amp) to thousands of amps. Ammeters and multimeters like the ones described in this lesson are not capable of measuring currents much above a few amps. For larger currents, it is impractical to build a series ammeter that can carry the full circuit current. This chapter is limited to a discussion of ammeters used for small DC currents.

Just as there are multiple scales for resistance on a multimeter, most multimeters also have multiple scales for use in current readings. Figure 6–6 shows an analog ammeter that has two DC current scales. The full-scale readings for these two scales are 0.5 DCmA and 25 DCmA. The maximum current that can be directly read using this meter is 25 mA. Currents greater than 25 mA can overdrive the meter element and damage it. With proper techniques, however, a shunt could be used with this meter to extend the range for current readings. You will learn more about shunts (a parallel path for current) in later chapters.

The scales for current on the analog meter are located at the center of the meter face. The scale nearest the bottom corresponds to the 0–0.5-DCmA scale and is calibrated in 0.01-DCmA increments. The scales used for current measurements are much more linear than the scale

FIGURE 6–5 Ammeter in a circuit.

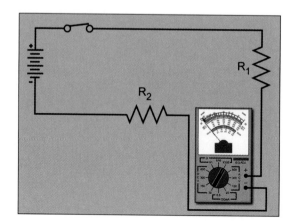

FIGURE 6–6   Analog ammeter.

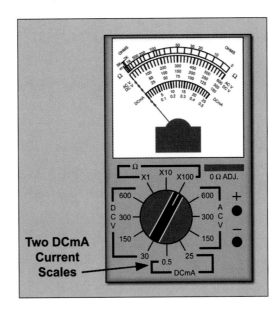

FIGURE 6–7   Ammeter circuit connection.

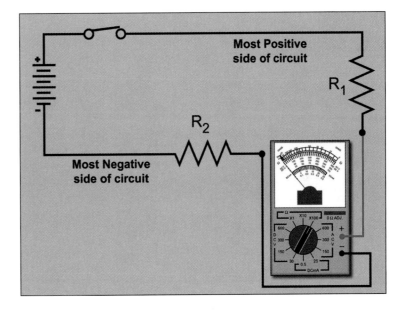

used earlier for resistance measurements. For this reason, it is much easier to read close to the ends of the scale. Maximum accuracy is still obtained, however, when the meter is read with the indicator near the center of the selected current scale.

When measuring current, you should always begin with the highest scale on the meter, then decrease the scale to obtain the proper reading if the selected scale is too high. Most modern electronic multimeters also have fuse protection on current scales to prevent internal damage to the meter. The presence or absence of such a meter protection fuse must be verified for each meter used.

When using an analog meter to measure DC current, the polarity of the current is indicated by the direction of the meter needle's deflection. Normally, the meter is connected so that the red test lead is connected to the more positive point in the circuit and the black test lead is connected to the more negative point in the circuit (see Figure 6–7).

FIGURE 6–8 Digital multimeter.

If the analog meter is connected incorrectly for more than a fraction of a second, the meter may be damaged. Damage to the meter movement (the part that moves the needle) may affect the accuracy of future meter readings.

CAUTION: When using any type of meter to measure electrical values in an energized circuit, be sure to follow the safety precautions recommended by the meter manufacturer. Do not attempt to use test equipment for purposes for which the meter was not designed. To do so will not only damage or destroy the meter but may also cause harm or personal injury to the individual performing the test.

Figure 6–8 shows a digital meter, which can be used to measure current. The digital meter includes five different current scales, including 200 μA, 2 mA, 20 mA, 200 mA, and 2 A. In addition, one position of the selector (20 m/10 A) serves a dual purpose. For all normal scales of the meter, the test leads are inserted into the "COM" and the "A" jacks of the meter. However, when the selector switch is set to the "20 m/10 A" position and the test leads are inserted into the "COM" and the "10 A" jacks, the meter will read 10 amps full scale. This is useful when measuring higher currents in a circuit. However, one additional precaution must be followed with the 10-amp scale. All the current scales for the digital meter are fuse protected with the exception of the 10-amp scale. If the current exceeds 2 amps for any of the current scales except the 10-amp scale, the internal (replaceable) fuse will blow, creating an opening to protect the circuit. If the current exceeds 10 amps when the meter is set up to read current on the 10-amp scale, the meter will be damaged or destroyed.

When the digital meter is used to measure current, the scale is selected using the meter's selector switch. The position of the decimal point on the digital display will indicate the range of readings for the current scale selected. Table 6–1 shows the scales for this meter.

An additional switch, one that must be set for the digital meter shown, is the DC/AC switch. This switch is used to tell the meter if the current being measured is DC or alternating current (AC). In this chapter, it is assumed that all currents are DC.

As with the resistance scales, an overcurrent or current larger than the selected scale would be displayed with a single "1" digit. Care must

Table 6–1   Digital Multimeter Display

Scale	Display at Maximum
200 μA	199.9
2 mA	1.999
20 mA	19.99
200 mA	199.9
2 A	1.999
10 A	10.00

be taken not to confuse the overcurrent indication (1., 1. , or 1.) with a valid reading such as "1.000," "1.00," or "100.0." Some digital meters indicate such an overload condition by displaying the letters "OL." Read the manual supplied with any meter you might use to find out how that meter indicates overcurrent or overload conditions.

When the current reading on a digital meter is underscale, the reading will be shown as a series of zeros with the decimal point correctly located for the scale being used. In the 200-mA scale position, for example, when there is no current flowing through the meter or the current through the meter is less than 100 mA, the meter display will be "00.0." If you see this reading and you suspect there should be some current present in the circuit, adjust the selector switch to the next lower range and try to measure the current. Repeat this procedure until the selected scale allows the circuit current to be measured or until you get to the lowest scale. Currents below 0.1 μA (microamps) are not detectable by this meter.

Digital meters can measure current of either polarity. When the red test lead is connected to the more positive point in the circuit and the black test lead connected to the more negative point, the digital meter will indicate the polarity by either showing a "+" sign ahead of the numerical reading or by not indicating a sign for the circuit current. When the red test lead is connected in the circuit to a point that is more negative than the black test lead, however, this reverse polarity will be indicated by a "−" sign ahead of the numerical reading of the digital display. The meter's accuracy is not affected by the polarity of the applied signal.

TIP: It is a good practice to always identify lead polarity and connect accordingly. This helps prevent misconnections when using an analog meter.

6.4 Measuring Current

When measuring current with either type of meter (analog or digital), the procedure to obtain accurate measurement of the current value is the same. As with resistance, digital meters are often more precise and easier to read for steady-state currents or currents that are not changing rapidly, while analog meters often work better when currents are changing or varying and will not settle down to a precise value.

The normal power or voltage supply for the circuit must be applied in order to take current readings. However, the circuit must be turned off in order to safely connect the ammeter before taking the current readings.

Caution must be used to ensure that the meter polarity is correct (even if using a digital meter) and that the scale selected for the reading is as great or greater than the highest anticipated circuit current. Generally, the meter's selector switch can be set to the highest scale and then adjusted downward if the scale is too high. In the case of the digital meter, if the current is within the limits of the 10-amp range but greater than the 2-amp range, then the "10 A" test lead jack and "20 m/10 A" switch selector position must be used. If later you wish to lower the range setting, power must be again removed from the circuit since removing the test lead will involve breaking the current path in the circuit.

Since there is a single current path through a series circuit, the circuit can be broken and the meter inserted at any place in the circuit. The current reading will be the same wherever the meter is inserted into the circuit. Following is a discussion of circuits in which it is necessary to measure the current.

Figure 6–9 shows the proper way to measure the current in this circuit. The circuit has been broken between resistors R_1 and R_2 and the meter inserted with the positive test lead toward the positive power supply lead. The meter could have been inserted at any place in the circuit by breaking the conductor at that location.

Figure 6–10 shows a circuit in which the current measurement would be in error because the meter is improperly installed. The meter must be placed in series with other circuit components not in parallel, as shown. If the meter were connected as shown, the internal resistance of the meter would be in parallel with resistor R_1, creating an additional current path and changing the parameters for the circuit, thus giving an erroneous reading for circuit current.

Figure 6–11 shows a circuit in which an analog meter is incorrectly connected. The negative meter lead is connected to the more positive circuit point and the positive meter lead to the more negative circuit point. This could damage the analog meter's movement. This connection would not be a problem, however, if a digital meter were used for this measurement.

There are "clamp-on" type ammeters that are used to measure circuit current without breaking the circuit. These work by sensing the magnetic field surrounding a conductor as a result of current flowing through that conductor. These meters are generally used on AC circuits.

FIGURE 6–9   Ammeter connected correctly.

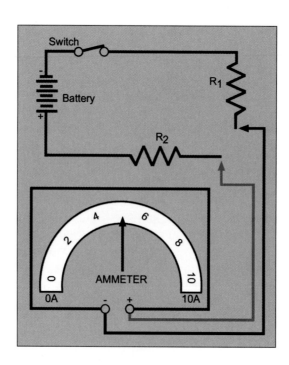

FIGURE 6–10 Ammeter
connected improperly.

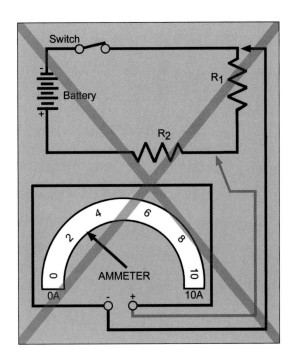

FIGURE 6–11 Ammeter
connected improperly.

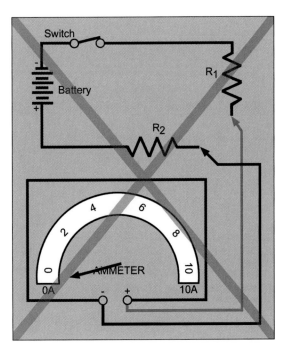

■ SUMMARY

In this chapter you were introduced to the schematic symbols for fuses and circuit breakers. You were also introduced to the fundamental operation of these important devices. Their operation is contingent on one of the fundamental principles of series circuits: The current in a series circuit is the same in every component.

In addition to being the same throughout the series circuit, the current flow is also directly proportional to the circuit supply voltage and inversely proportional to the total circuit resistance. This means simply that if the voltage goes up, the current goes up by the same percentage. If the resistance in the circuit goes up, the current goes down by the same percentage. Or, as stated in Ohm's law, $I_T = E_R/R_T$.

Clearly, if one or more resistors or components in a circuit are changed in resistance value, the circuit current will increase or decrease accordingly. Of course, the current is a function of the total resistance and the total voltage; however, current can be calculated using the voltage across one component and its resistance. This is true because the current has to be the same throughout the series circuit and can be calculated at any component.

Measurement of current in a circuit is as straightforward as measurement of resistance. In current measurements, however, the ammeter (either digital or analog) is placed in series with the circuit. After the connection polarities are confirmed and the scales properly ranged, the circuit can be energized and the current read on the meter.

■ REVIEW QUESTIONS

1. The current in a series circuit is the same in every component. Discuss the meaning of this and what effect it might have on your work as an electrician.

2. Both fuses and circuit breakers are used to protect an electrical circuit. How do they do this, and what differences are there between them?

3. Depending on what information and equipment you have available, there are several ways that the current can be determined in a series circuit. Discuss three of these methods.

4. A certain circuit has a total resistance of 50 ohms with a supply voltage of 10 volts. What will happen to the current in the circuit if the resistance is cut in half? What will happen if the voltage is doubled?

5. Compare and contrast the digital ammeter versus the analog ammeter. Which is more accurate? Which might be more useful in a circuit with varying current? Which is most sensitive to polarity?

■ PRACTICE PROBLEMS

1. In the table that follows, fill in the missing information.

E_T (volts)	R_1 (Ω)	R_2 (Ω)	VR_1 (volts)	VR_2 (volts)	I_T (amperes)
	6	10	12		
120	5	5			
60	10				3
25		2		10	
75	10			35	

2. A certain circuit has two resistors of equal value. What will happen if both the resistors are doubled in resistance? What will happen if one of the resistors is doubled in value?

3. Assume that the fuse in Figure 6–4a is a 0.25-ampere fuse. What will happen to that fuse if a short circuit occurs across resistor R_1?

4. In Figure 6–7, assume that the meter is in the 0.5-DCmA range. If the needle reads halfway between 0.2 and 0.3, how much current is flowing in the circuit?

5. You are going to measure DC current in a circuit using the digital multimeter shown in Figure 6–8. You expect the current to be 4.5 amperes. Where should the leads be connected to the meter?

chapter 7

How Voltage Functions in a DC Series Circuit

◼ OUTLINE

■ OVERVIEW

This chapter will finish the sections on how resistance, current, and voltage function in a series circuit. In chapter 6, you learned that the current is equal in all components of a series circuit. Here you will learn that the sum of all the component voltages must be equal to the supply voltage. These concepts are very important from the standpoint of the practicing electrician and his or her day-to-day work. The information in chapters 5 to 7 will eventually be combined to provide the background for more advanced circuit analysis in series, parallel, and combination circuits.

■ OBJECTIVES

After completing this chapter, you should be able to:

1. Draw and label components of series electrical circuits including single- and multiple-voltage sources.
2. Calculate the effective voltage applied to series circuits.
3. Use Ohm's law to determine the voltage applied to a series circuit or to the individual components in a series circuit.
4. Measure the voltage across the components of a series circuit using either an analog or a digital voltmeter.

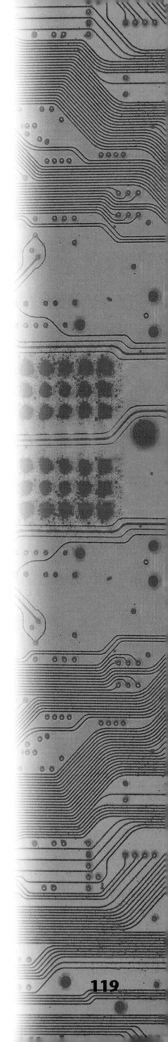

▪ VOLTAGE EVALUATION AND CALCULATION IN SERIES CIRCUITS

7.1 Introduction

Every electrical circuit must have a voltage source in order to operate. The voltage source provides potential energy to the circuit. The circuit uses this energy to accomplish work. No work is done in the circuit unless the voltage source is connected to the circuit in such a way that current flows through the circuit. Of course, there are voltages throughout the circuit. Voltages that are caused by currents flowing through resistances are referred to as *voltage drops.*

In a series circuit, voltage sources are connected in series with the load or loads. This circuit voltage supply may be a simple voltage source, such as a single battery, or it may be more complex, such as several voltage sources connected in series or parallel. For example, some flashlights require more than one battery. Another possible source of voltage for a series circuit might be a generator, a machine that converts mechanical energy to electrical energy. Generators will be discussed in detail later.

Regardless of how the voltage is supplied in a series circuit, it is the source that determines the energy available to that circuit. Circuit voltage is as important in determining circuit current as the total resistance.

In this chapter, the emphasis will be on the voltage source found in the circuit and on the voltages developed across individual components in the circuit. Unlike current, the voltage across each component in a series circuit will vary in proportion to the resistance of the component (the higher the resistance of the component, the higher the voltage drop across the component).

7.2 Voltage Sources

So far, you have studied circuits containing one simple battery. What if the circuit requires a larger voltage source than is available from a single battery? One circuit that almost always requires a larger voltage than is available in an individual battery is a household flashlight. A common flashlight requires two D-cell batteries (see Figure 7–1). Each battery is 1.5 VDC. This means that the lamp in the flashlight is a 3-volt light.

Figure 7–2 shows the schematic equivalent of two 1.5-VDC batteries connected in series. Notice that the negative terminal of one battery is connected to the positive terminal of the next battery. This causes the voltage of one battery to be added to the next. This is called a *series-additive power source.* As the voltmeter shows, the total voltage is 3.0 VDC across both batteries. In other words, the total voltage is the sum of the voltages added in series, regardless of their values:

$$E_T = E_1 + E_2 + E_3 + \dots E_n \qquad (1)$$

If batteries are connected in series but are not connected positive terminal to negative terminal, we call them *series-opposing power sources.* Figure 7–3 is an example of series-opposing batteries. In this case, the total voltage is equal to the larger voltage minus the smaller voltage:

$$E_T = E_{Larger} - E_{Smaller}$$

FIGURE 7–1   Flashlight batteries can be connected in series to increase the total voltage.

FIGURE 7–2   Batteries connected in series-additive configuration.

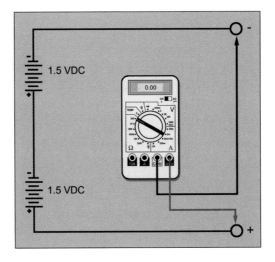

FIGURE 7–3   Series-opposing power sources.

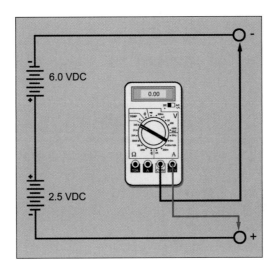

Figure 7–4 is a combination of series-additive and series-opposing power sources. Note that in this circuit, two of the voltages are connected in one direction or polarity, and the other two voltages, while matching each other, are connected in the opposite direction relative to the first group. To solve for the total resulting voltage from all the sources, the voltage sources must be grouped by polarity and then each group added separately. In the circuit shown, the 3.0 V and 6.0 V are both the same polarity, while the 1.5 V and the other 1.5 V both have the same polarity but are connected in the opposite direction relative to the first group. These two groups must be added, and then the smaller value must be subtracted from the larger. The work is shown here:

Group 1:

$$E_{\text{Group 1}} = E_1 + E_4$$
$$E_{\text{Group 1}} = 1.5 \text{ V} + 1.5 \text{ V}$$
$$E_{\text{Group 1}} = 3.0 \text{ V}$$

FIGURE 7–4   Series-adding and series-opposing voltages.

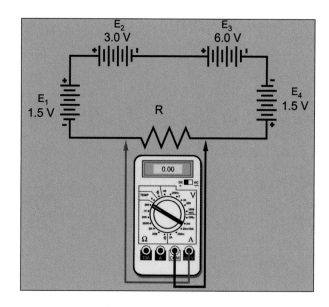

Group 2:

$$E_{\text{Group 2}} = E_2 + E_3$$

$$E_{\text{Group 2}} = 3.0 \text{ V} + 6.0 \text{ V}$$

$$E_{\text{Group 2}} = 9.0 \text{ V}$$

$$E_{\text{TOT}} = E_{\text{Larger Group}} - E_{\text{Smaller Group}}$$

$$E_{\text{TOT}} = E_{\text{Group 2}} - E_{\text{Group 1}}$$

$$E_{\text{TOT}} = 9.0 \text{ V} - 3.0 \text{ V}$$

$$E_{\text{TOT}} = 6.0$$

This shows, by adding the various voltages in groups and then subtracting the smaller group total from the larger group total, that the total amount of resulting voltage supplied to the circuit from all the various voltage sources can be calculated.

7.3 Voltage Drop across Components

Now that you have learned about voltages in series sources, look at what happens to each individual component in a circuit. To do this, apply Ohm's law for voltage and current:

$$E = I \times R \text{ and } I = \frac{E}{R} \text{ and } R = \frac{E}{I}$$

EXAMPLE 1

Using Figure 7–5, determine the voltage drop across each resistor.

Solution:
To determine the voltage across each of the resistors (assume that the fuse, wire, and switch have no resistance), the current through each resistor must be determined. Since this is a series circuit, the current

FIGURE 7–5   Series voltages.

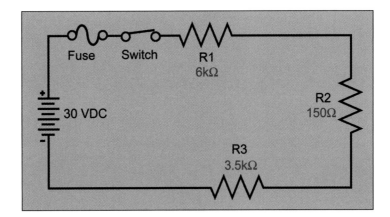

through any component is the same as any other component. The first thing we must do is calculate the total resistance in the circuit. So,

$$R_T = R_1 + R_2 + R_3$$
$$R_T = 6 \text{ k}\Omega + 150 \text{ }\Omega + 3.5 \text{ k}\Omega$$
$$R_T = 6{,}000 \text{ }\Omega + 150 \text{ }\Omega + 3{,}500 \text{ }\Omega$$
$$R_T = 9{,}650 \text{ }\Omega$$
$$I_T = I_1 = I_2 = I_3$$

and

$$I_T = \frac{E_T}{R_T}$$

So,

$$I_T = 30 \text{ V}/9{,}650 \text{ }\Omega$$

$$I_T = 3.11 \text{ mA or }.00311 \text{ A}$$

This means that 3.11 mA are flowing through each resistor in the circuit. To determine the voltage across each resistor, use Ohm's law again to get

$$E = I \times R$$
$$E_1 = (0.00311)(6{,}000)$$
$$E_1 = 18.66 \text{ V}$$
$$E_2 = (0.00311)(150)$$
$$E_2 = 0.46 \text{ V}$$
$$E_3 = (0.00311)(3{,}500)$$
$$E_3 = 10.88 \text{ V}$$

These may seem like some strange answers, but look at the results. The resistor with the largest resistance has the largest voltage drop. This makes sense since it is providing a larger portion of the total resistance. In addition, the resistor with the smallest resistance drops the least amount of voltage.

Now, add all the resulting voltages:

$$E_T = E_1 + E_2 + E_3$$
$$E_T = 18.66 \text{ V} + 0.46 \text{ V} + 10.88 \text{ V}$$
$$E_T = 30 \text{ VDC}$$

This shows that the sum of all the voltages dropped across each of the circuit components is equal to the source.

EXAMPLE 2

What would happen to the voltage drop across each component if the source voltage in Figure 7–5 increased to double its original value?

Solution:
Since the voltage has changed, the total current must be determined:

$$I_T = \frac{E_T}{R_T}$$
$$I_T = \frac{60 \text{ V}}{9,650 \ \Omega}$$
$$I_T = 6.22 \text{ mA or } .00622 \text{ A}$$

and

$$E = I \times R$$
$$E_1 = (0.00622)(6,000)$$
$$E_1 = 37.32 \text{ V}$$
$$E_2 = (0.00622)(150)$$
$$E_2 = 0.93 \text{ V}$$
$$E_3 = (0.00622)(3,500)$$
$$E_3 = 21.77 \text{ V}$$

Example 2 shows that an increase in the applied voltage causes the voltage drop across each component to increase by the same factor: The voltage source doubled, and the component voltage drop doubled.

With the information you have studied so far, it can be seen that components in series having the same resistance will have the same voltage drop. The proof is this: The total current is the same as the current through any single component, and components having equal resistance rating will drop equal voltages ($E = I \times R$). The following calculations, based on the circuit in Figure 7–6, proves that these voltage drops are equal.

Since the current is the same throughout the series circuit, to solve for the current, the total resistance must be calculated:

$$R_T = R_1 + R_2 + R_3$$
$$R_T = 5 \text{ k}\Omega + 5 \text{ k}\Omega + 3.5 \text{ k}\Omega$$
$$R_T = 5,000 \ \Omega + 5,000 \ \Omega + 3,500 \ \Omega$$
$$R_T = 13,500 \ \Omega$$

FIGURE 7–6   Series circuits.

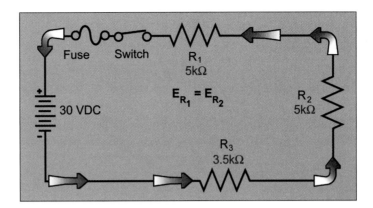

Using the previous calculation, the current can be calculated as follows:

$$I_T = \frac{E_T}{R_T} = \frac{30}{13,500}$$

$$I_T = .0022 \text{ A or } 2.2 \text{ mA}$$

Now that the circuit current has been calculated and you know that the current is the same throughout a series circuit, you can prove that the voltage drop across resistors of the same value will also be the same:

$$E_1 = I_1 \times R_1$$
$$E_1 = .0022 \times 5,000$$
$$E_1 = 11 \text{ V}$$

$$E_2 = I_T \times R_2$$
$$E_2 = .0022 \times 5,000$$
$$E_2 = 11 \text{ V}$$

This proves that when the resistances are equal in a series circuit, the voltage drop across those resistances is also equal.

7.4 Direction of Current Flow

Until now, not much attention has been paid to the direction of current flow. But current flow has a specific direction in a series DC circuit. In Figure 7–6, you'll notice the battery has a positive and a negative side. The negative terminal is labeled "negative" because that is where there is a buildup or excess of electrons in the battery. Since electrons have a negative charge, this terminal is labeled negative. The excess electrons leave the negative terminal and travel through each circuit component and back to the positive terminal. So, outside the battery, the current flows from negative to positive.

■ USING A VOLTMETER

7.5 Voltmeter Types and Construction

In previous chapters you learned how to use a multimeter to measure both resistance and current in a series circuit. In this section, you will learn how to use the same test equipment to measure the voltage across any two points in a series circuit. It takes two points because the volt-meter measures the difference in potential between the two points. Both meters discussed previously contain scales for measuring voltage. Unlike the procedure used for current measurement, the meter does not become part of the circuit's current path when measuring voltage.

This is not to say the internal resistance of the meter does not affect the circuit's operation, only that the voltmeter is not supposed to alter the operation of the circuit. Part of the circuit current is bypassed through the voltmeter to obtain a voltage reading. The amount of current flowing through a good, high-impedance meter is negligible and does not affect the operation of the circuit.

The voltmeter can be used to measure the voltage across any device in a circuit. Voltage is the driving force that produces movement of electrons in a circuit. Ohm's law relates all voltage, current, and resistance in a circuit. The amount of current in a circuit with a fixed resistance or load will be directly proportional to the amount of voltage present. Voltage may exist, however, even though no current is flowing in the circuit. A voltmeter can be used, for example, to measure the potential in a battery, even though the battery is not connected in a circuit and no current is flowing from the battery.[1] Voltage potential can exist either in or out of a circuit. Current is present only when both a voltage source and a current path are present.

As with resistance and current, voltages can vary from very small values (millivolts and microvolts) to very large values (megavolts). The meters described in this chapter can measure voltages as low as a few millivolts or, depending on the type of meter and the type of voltage being measured, as high as 1-kilovolt, or 1,000 volts. For safety purposes, it is important to check the maximum voltage rating of a multimeter before using. Many multimeters have a maximum rating of 600 volts. To accommodate this range of voltages, different voltage scales are included with each meter.

Figure 7–7 shows the analog meter discussed in earlier lessons. The voltage scales are located between the resistance and current scales on this meter. Three different numerical scales are provided for the voltage ranges on the analog meter in order to provide four different DC voltage ranges. The DCV ranges included on the analog meter include 0–30, 0–150, 0–300 and 0–600 volts. The same faceplate scales are used for both the DC and the AC measurements. However, different selector switch positions are used. The markings for the various scales are broken into 1-volt, 5-volt, 10-volt, and 20-volt increments, depending on the scale being used.

When measuring voltage, you should always begin with the highest scale. You can then reduce the scale using the selector to give the most accurate reading. Generally, voltmeters are temporarily connected to a circuit using test probes, but voltmeters of a nonmultimeter type can be installed where a continuous monitoring of a circuit voltage is required.

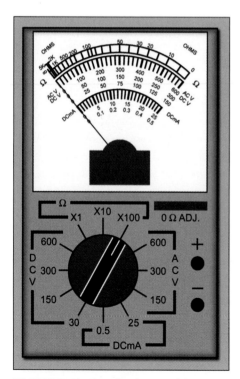
FIGURE 7–7 Analog multimeter.

[1]Actually a very small current does flow into the voltmeter; however, it is small enough that no significant change occurs in the circuit.

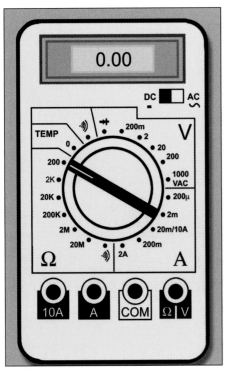

FIGURE 7–8   Digital multimeter.

Voltages are polarized just like currents. When using the analog meter, you must connect the meter to the correct polarity. The red test lead is inserted into the positive (+) test-lead jack and is connected to the more positive point in the circuit. The black test lead is inserted into the negative (−) test-lead jack and is connected to the more negative point in the circuit. As in performing current tests, the meter may be damaged if left connected to the wrong polarity for more than a fraction of a second.

Figure 7–8 shows the digital meter discussed in earlier lessons. The digital meter has five different voltage ranges as well as a special voltage scale indicated by the diode symbol (⊣⊢). Voltage ranges on the digital meter include 200 millivolts and 2, 20, 200, and 1,000 volts. The special scale that is designated by the diode symbol is a special voltage scale used for testing semiconductor devices and is slightly different from a normal voltage scale. This special scale is common but is not important for the discussion at this time.

To measure voltage using the digital meter, the test leads are inserted in the test-lead jacks marked "COM" and "Ω/V." When used to measure voltage, the scale selected on the meter's selector switch will determine the position of the decimal point on the display. Scales for this meter are shown in Table 7–1. As in current measurements, the DC/AC switch must be set to select with DC or AC voltage measurements.

Overvoltages would be displayed as a "1" and should not be confused with voltage displays where all digits would be shown. Some meters display the symbol "OL" (overload) to show voltage values that are beyond the range of the meter. Be sure to read the manual supplied with the meter you might use to find out how it indicates overvoltage or overload conditions.

When the voltage reading on a digital meter is underscale, the display will show a series of zeros with a decimal point correctly located for the scale being used. In the 200-mV position, for example, when there is no voltage present at the test leads or the voltage is less than 0.1 mV, the meter display will be "00.0." If you get such a display on any scale other than the 200-mV scale (already the lowest scale available) and you suspect there should be some voltage present at the test leads, adjust the selector switch to the next lower range to attempt to measure the voltage. Repeat this procedure until the selected scale allows the voltage to be measured or until you reach the lowest scale. Voltages below 0.1 mV cannot be measured using this multimeter.

Table 7–1   Digital Multimeter Display

Scale	Display at Maximum
200 mV	199.9
2 V	1.999
20 V	19.99
200 V	199.9
1,000 V	1.000

Digital meters can measure voltage of either polarity. When the red test lead is inserted into the "Ω/V" test-lead jack and connected to the more positive point and the black test lead is inserted in the "COM" test-lead jack and connected to the more negative point, the meter will indicate a positive voltage by displaying a "+" or by displaying no sign ahead of the numerical value. If the polarity of the test leads is reversed, the numeric display will be preceded by a "−" sign. The digital meter's accuracy is not affected by the polarity of the applied voltage.

7.6 Measuring Voltage

When measuring voltage with either type of meter, the procedures for measuring voltage are the same. As with resistance and current, voltage values are more easily read with digital meters when the voltages are steady state or not changing. Generally, analog meters work better for voltage values that fluctuate rapidly or that will not settle down to a precise value.

Most voltmeters measure voltage by drawing a very small amount of current from the circuit being tested. The voltage to be measured is used to produce a small current through a known resistance. The meter is then calibrated to show the voltage rather than the current. All voltmeters produce some error when connected to a circuit, but some designs produce less error than others. Generally, the higher the input impedance or load (resistance) provided by the meter, the more accurate the voltage reading will be. Meters with very high input impedances (greater than 10 million ohms) generally do not affect the circuit to which they are connected since the current through the meter is extremely small (see Figure 7–9). Each range creates a preset voltage drop across the meter movement to read the voltage being measured.

Analog meters are polarity sensitive and may be damaged if connected incorrectly. Be sure the scale selected for the voltage measurement is as great as or greater than the highest anticipated circuit voltage. Generally, the meter's switch should be set to the highest voltage scale and then adjusted downward if the scale is too high.

In electrical and electronics circuits, voltages are developed or are present across each component in the circuit. Each of these voltages

FIGURE 7–9 Typical voltmeter circuit for multiple ranges.

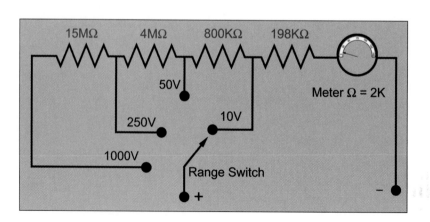

can be measured independently, or the voltage across two or more components can be measured at the same time. Voltage measurements are measurements of the difference in potential between any two points in a circuit.

Figure 7–10 shows the proper way to measure the voltage across resistor R_1. The voltmeter is connected across the resistor and will show only the voltage that is developed as current flows through that resistor. The positive meter lead is connected to the more positive end of the resistor (the end closest to the positive terminal of the power supply or battery), and the negative test lead is connected to the more negative end of the resistor (the end closest to the negative terminal of the power supply or battery).

Figure 7–11 shows a voltmeter connected across both resistors R_1 and R_2. The voltage indicated would be the total voltage developed across both of these resistors, which would be equal to the power supply voltage.

Figure 7–12 shows a voltmeter inserted in series with the circuit between resistors R_1 and R_2 as it might be connected to measure current. Ammeters have a very low internal resistance and do not significantly affect the circuit when installed in series. Voltmeters, as discussed earlier, have a very large internal resistance. As a result, the current path would be greatly altered if the voltmeter were installed in this manner. The voltmeter reading would not reflect the voltage across R_1 or R_2 but would indicate a voltage reading very close to the power supply potential since the very small current flowing through the meter would not produce a significant voltage drop across resistors R_1 and R_2.

You can prove this by using Ohm's law. In the circuit shown in Figure 7–12 assume the resistance value of R_1 is 5,000 ohms and the resistance value of R_2 is 10,000 ohms. Since a multimeter has a very

FIGURE 7–10 Voltmeter connected properly.

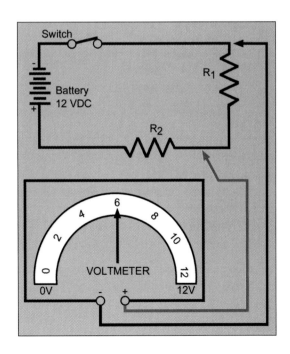

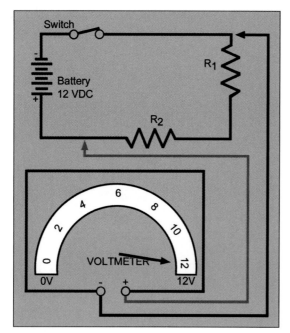

FIGURE 7–11   Voltmeter connected across both resistors (correct).

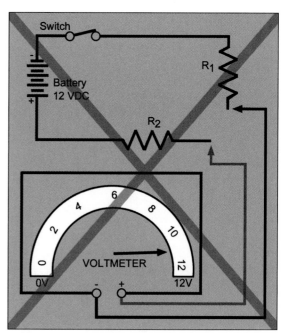

FIGURE 7–12   Voltmeter connected in series with the circuit (incorrect).

high resistance when it is used to measure voltage, assume a resistance value for the multimeter of 5,000,000 ohms:

$$I_T = \frac{E_T}{R_T} = \frac{12}{(5\ k\Omega + 10\ k\Omega + 5\ M\Omega)}$$

$$I_T = 2.39\ \mu A$$

Voltage drop across the meter = 2.39 μA × 5 MΩ

Voltage drop across the meter = 11.96 V

As can be seen by the previous calculation, because of the high internal resistance of the multimeter when used to measure voltage, the value of the reading when the meter is placed in series is almost equal to the source voltage.

■ SUMMARY

In this chapter you completed your study of the major fundamentals of DC series circuits by examining voltage behavior. Voltage sources in series circuits may also be connected in series. Thus, you may have several flashlight batteries in series to provide a greater voltage, power, and light.

When voltage sources are connected in series, the total voltage is equal to the sum of each individual source; however, you must take polarity into account. When two voltage sources are connected in series with the positive terminal of one connected to the negative terminal of the other, the total voltage applied to the

circuit will be the difference between the two sources. The larger voltage source will determine the polarity.

Once you have determined the total voltage in the circuit, the voltages across any individual element and the total current can be determined by using Ohm's law.

Sometimes it is necessary to measure the voltage in a circuit or across components of the circuit. The key elements to remember when measuring voltage are to be sure that the voltmeter is properly connected in parallel with the component and that the voltmeter is on the proper range and (especially in the case of an analog meter) connected with the proper polarity.

■ REVIEW QUESTIONS

1. What is the pressure that is applied to a circuit called?

2. There are two resistors in a circuit, one 50 ohms and one 100 ohms. Which resistor has the most voltage drop across it?

3. What is the formula for adding voltage in a series circuit?

4. How would you normally calculate a voltage drop across a resistor in a series circuit?

5. When using a multimeter to measure voltage, should you always start from the lowest scale and work you way up? Why or why not?

6. When measuring fluctuating voltages, what meter is usually better suited: the analog or the digital meter? Which might be more useful in a circuit with steady voltages? Which is most sensitive to polarity? Why?

■ PRACTICE PROBLEMS

1. A series circuit has three resistors. R_1 has a voltage drop of 10 volts, R_2 a voltage drop of 20 volts, and R_3 a voltage drop of 30 volts. What is the total voltage of this circuit? Draw the circuit.

   a. 20
   b. 40
   c. 60
   d. 80

2. A series circuit has three lamps. Lamp 1 has a 15-ohm resistance, lamp 2 has a 20-ohm resistance, and lamp 3 has a 25-ohm resistance. The circuit has a total current of 2A.

   a. What is the voltage drop across each lamp?
   b. What is the wattage of each lamp?

3. A series circuit has three resistors with the following values:

Resistor	Resistance	Voltage
R_1	20	40
R_2	30	60
R_3	??	

The total voltage applied to all three resistors is 120 volts.

   a. What is the voltage across R_3?
   b. What is the current through the circuit?

c. What is the resistance of R_3?

d. What is the power dissipated by each resistor and the total power?

4. Consider this circuit:

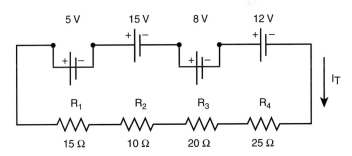

Fill in the following information:

$I_T = $ _____

$V_T = $ _____

Resistor	Voltage	Watts
R_1		
R_2		
R_3		
R_4		

chapter 8

How Voltage Dividers Work in a DC Series Circuit

■ OUTLINE

INTRODUCTION   LAW OF PROPORTIONALITY
VOLTAGE DIVIDERS

OVERVIEW

In this chapter you will learn one of the most useful circuit concepts in electricity: the ability to take a voltage supply and create multiple voltages using only resistors. This concept, called a *voltage divider,* is a frequently used circuit whether you are looking for an extra voltage to supply power, analyzing a circuit to troubleshoot, or modifying a circuit so that it works more appropriately for the task at hand.

The skills you learn in this chapter, while seemingly simple, will be used again and again in your electrical career.

OBJECTIVES

After completing this chapter, you should be able to:

1. State the law of proportion for series circuits.
2. Describe voltage dividers in series circuits.

GLOSSARY

Voltage divider An electrical circuit, usually made up of resistive elements, that can be used to break down, or divide, a supply voltage to two or more smaller voltages.

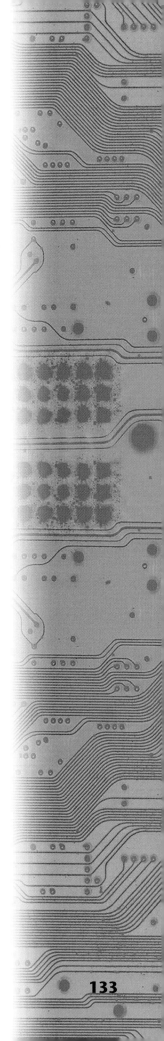

■ INTRODUCTION

In previous chapters, Ohm's law has been used exclusively to calculate resistance, current, and voltage values for resistors in series DC circuits. The Law of Proportionality also uses Ohm's law to find the voltage across a single resistor in a series of resistors. This law describes the relationship among resistors and circuit resistive elements in a circuit. Using this law, you can calculate the voltage across any resistor or combination of resistors in a series circuit without knowing the circuit current.

This chapter also covers the method used to separate circuit voltages into different values. These different voltages may be required by other parts of the same circuit or by other circuits. Circuits of this type, where the voltage is divided between two or more resistors, are called **voltage dividers**. They "divide" the voltage into as many different values as are needed for different applications.

■ VOLTAGE DIVIDERS

Refer to the simple voltage divider circuit in Figure 8–1. This is a voltage divider with two resistors (R_1 and R_2) and a DC voltage source. The voltage applied is 30 VDC. Assume that the resistance values of both R_1 and R_2 are equal to each other ($R_1 = R_2$). The following has been shown from previous lessons:

1. The current through a DC series circuit is the same value through all the components.
2. The total resistance value of a DC series circuit is the sum of all the resistance values in the circuit.
3. The sum of all the voltage drops in a DC series circuit is equal to the source voltage.

Since both R_1 and R_2 are of equal resistances, the voltage drop across each one will be the same number. This is shown in the following equations:

Since	$R_1 = R_2$	(given)
and	$I_1 = I_2 = I_T$	(series circuit)
then	$I_1 \times R_1 = I_2 \times R_2$	
so	$E_1 = E_2$	(Ohm's law)

FIGURE 8–1 Simple voltage divider.

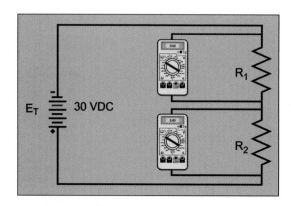

In chapter 7, you learned that the sum of the voltages in a series circuit is equal to the voltage source (E_T). Since $E_T = E_1 + E_2$ in Figure 8–1 and $E_1 = E_2$, E_1 and E_2 are each half the amount of E_T. The voltage divider in Figure 8–1 divided the total voltage into two equal parts.

The voltage divider in Figure 8–1 is relatively simple since both resistors are of equal value. The following description covers a circuit whose resistors are not equal. R_1 in Figure 8–2 is 10 Ω, and R_2 is 5 Ω—R_1 is twice the value of R_2. Using the statements above and Ohm's law, we can show how the voltage drops across the resistors compare to one another:

Since	$R_1 = 2R_2$	(given)
and	$I_1 = I_2$	(series circuit)
	$\dfrac{E_1}{R_1} = \dfrac{E_2}{R_2}$	(Ohm's law)
	$E_1 = \dfrac{E_2 \times R_1}{R_2}$	
and since	$R_1 = 2R_2$	
then	$E_1 = \dfrac{E_2 \times 2R_2}{R_2}$	(substitution)
so	$E_1 = 2E_2$	(canceling the R_2)

The sum of the voltages across all the resistors is equal to the source voltage (E_T):

$$E_T = E_1 + E_2$$

Since
$$E_1 = 2E_2$$

then
$$E_T = 2E_2 + E_2$$

$$E_T = 3E_2$$

so
$$E_2 = 1/3E_T$$

Comparing this to the resistances,

$$R_T = R_1 + R_2$$

$$R_T = 2R_2 + R_2$$

$$R_T = 3R_2 \text{ (substitution)}$$

$$R_2 = 1/3R_T$$

FIGURE 8–2   Analyzing a two-resistor voltage divider.

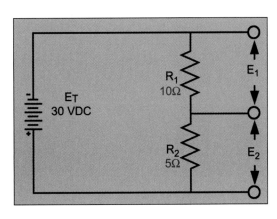

This shows that the voltage dropped across any resistor is equal to the fraction that resistor is of the total resistance in a series circuit. R_2 is one-third of the total resistance, so it dropped one-third of the total voltage.

■ LAW OF PROPORTIONALITY

Since the current through a series circuit is the same at any point, the voltage across any resistor is

$$E_R = I_T \times R_R$$

where E_R is the voltage across the resistor being measured, I_T is the total current, and R_R is the resistance of the resistor being measured. The total current in the circuit is represented by

$$I_T = \frac{E_T}{R_T}$$

Now, if the equation for total current is placed in the previous equation for the voltage across any resistor, the previous equation becomes Equation 1:

$$E_R = E_T\left(\frac{R_R}{R_T}\right) \tag{1}$$

Equation 1 is known as the Law of Proportionality. It says that the voltage dropped across any resistor in a series circuit is equal to the total voltage multiplied by the fraction of the value of the resistor in question divided by the total resistance.

The Law of Proportionality can be extended to calculating the voltage drop across more than one resistor. Consider Figure 8–3, for exam-

FIGURE 8–3 Analyzing a three-resistor voltage divider.

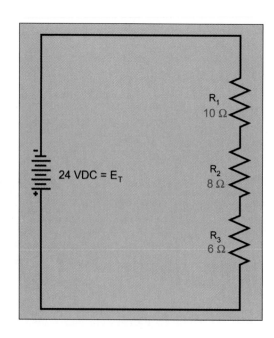

ple. If you wish to calculate the total voltage drop across resistors R_2 and R_3, you can do so by substituting into Equation 1 with the sum $R_R = R_2 + R_3$. This means that the total voltage drop across R_2 and R_3 can be calculated as

$$E_{R_2 + R_3} = E_T\left(\frac{8 + 6}{8 + 6 + 10}\right) = 24 \times \frac{14}{24} = 14 \text{ volts}$$

You can verify this by inspection of Figure 8–3.

■ SUMMARY

In this chapter you learned the principles of the voltage divider circuit. Voltage dividers are used throughout electricity and electronics to provide multiple voltages. The law of proportionality voltage allows you to calculate outputs from voltage dividers without knowing the current magnitude.

Voltage dividers can be used to provide multiple voltages from single voltage power supplies, provide selected voltages for biasing semiconductor circuits, set proportional voltage outputs for metering circuits, and a host of other uses.

Also, voltage divider concepts are a very powerful tool that can be used in the analysis of complex circuits.

■ REVIEW QUESTIONS

1. Voltage dividers in a series circuit may be calculated using Ohm's law or what else?
2. State the formula for finding the total resistance in a series circuit.
3. State the formula for finding total voltage in a series circuit.
4. State the formula for current in a series circuit.
5. State the formula known as the Law of Proportionality.

■ PRACTICE PROBLEMS

1. Using the equation for the Law of Proportionality, calculate the voltage drop across each resistor in Figure 8–3.

   You can answer the question by filling in the blanks in the following table:

$$E_1 = E_T \times \frac{R_1}{R_T} \qquad E_2 = E_T \times \frac{R_2}{R_T} \qquad E_3 = E_T \times \frac{R_3}{R_T}$$

$$E_1 = (\quad \text{VDC})\left(\frac{\quad \Omega}{\quad \Omega}\right) \quad E_2 = (\quad \text{VDC})\left(\frac{\quad \Omega}{\quad \Omega}\right) \quad E_3 = (\quad \text{VDC})\left(\frac{\quad \Omega}{\quad \Omega}\right)$$

$$E_1 = \quad \text{VDC} \qquad E_1 = \quad \text{VDC} \qquad E_1 = \quad \text{VDC}$$

2. Fill in the missing numbers in the following figure:

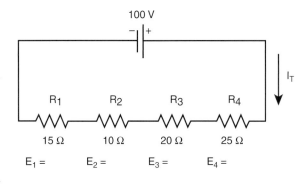

chapter 9

How to Calculate Power in a DC Series Circuit

■ **OUTLINE**

■ OVERVIEW

In earlier chapters you learned how to calculate resistance, current, and voltages in series circuits. You also had some limited exercises in calculated power. This chapter will add more information to your knowledge of power calculations and will teach you how to determine power requirements for elements used in a series circuit. As you study this material, remember that the things you learn in this chapter along with all the other material are key to your long-term success as an electrician.

■ OBJECTIVES

After completing this chapter, you should be able to:

1. Draw and label series electrical circuits that contain power-rated devices.
2. Calculate the total power used in series circuits.
3. Calculate the power used by individual components in a series circuit.

■ GLOSSARY

Ampere-hours The energy rating for a battery (see text for a definition).

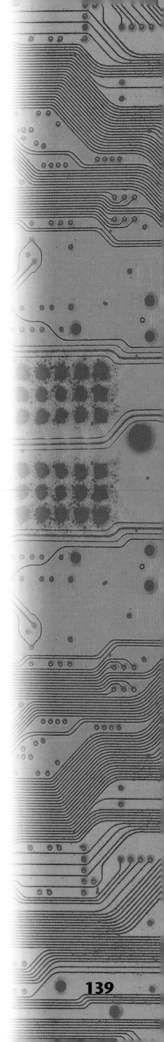

■ REVIEW OF ENERGY AND POWER

9.1 Energy

Remember that work and energy are essentially the same thing and are measured in the same units. For example, work is performed if you move a 200-pound crate a distance of 10 feet. The work performed is calculated as (200 pounds × 10 feet) = 2,000 foot-pounds, no matter how long it takes to move the crate the required distance. After the work is performed, you say that you have expended 2,000 foot-pounds of energy.

Electrical energy is measured in joules. The joule represents the amount of energy required to move 1 coulomb of charge through 1 volt of potential, or 1 joule = 1 volt-coulomb.

Other important energy unit is the British Thermal Unit (BTU). One BTU is the amount of energy required to raise the temperature of 1 pound of water 1 degree Fahrenheit. BTUs are often used in the rating of air conditioning equipment.

The electron volt (eV) is the unit of work used for electrons. Note that the electron is a charge and that voltage is a potential difference. One electron volt is the amount of work required to move an electron between two points with a potential difference of 1 volt. Since we know that 1 coulomb = 6.25×10^{18} electrons and that 1 joule = 1 volt-coulomb, then $1 J = 6.25 \times 10^{18}$ eV. The electron volt is used primarily in physics.

9.2 Power

Power is defined as the rate at which work is being performed, or, put another way, power is the rate at which energy is being expended. For example, the power used in moving the 200-pound crate mentioned earlier is equal to the work performed divided by the time required to move the crate. If it took 1 second to move the crate, then the power would be 2,000 foot-pounds per second. If it took 10 seconds to move the crate, then the power would be 200 foot-pounds per second.

Shown mathematically, this means that $\text{Power} = \dfrac{\text{Work}}{\text{Time}}$, or Work = Power × Time.

In a similar way, electrical power is the rate at which an electrical charge is forced to move by voltage. Remember that electrical charge in motion is called *electrical current,* and electrical current carries the energy from one location in the circuit to another. Electric power is measured in watts, which is defined as 1 watt = 1 joule/second. Since we know (from the definition of a joule) that 1 joule = 1 volt-coulomb and 1 ampere = 1 coulomb per second, this means that 1 watt = 1 volt × 1 ampere.

Because of the large energy requirements of modern electrical systems, electrical power is often expressed in thousands of watts, or kilowatts. Even larger systems may use millions of watts or megawatts. The unit of watts was selected in honor of the English scientist James Watt (1736–1819), who did much early research in the development of the theories concerning steam engines. It was Watt who realized that people of his era would better understand the amount of power being generated by equipment if it was related to the horse since everyone was familiar with the horse. He found that the average horse could lift 550 pounds, 1 foot in height, in 1 second.

Even in modern times, most motors are rated in horsepower. Remember that motor horsepower is the mechanical power capability of the motor. The amount of power into the motor (in watts or kilowatts) will be greater than the motor horsepower because of the energy losses due to motor resistance, windage, and friction.

Battery energy capabilities are specified in terms of the number of amperes they will supply for a given time. For example, if a battery will supply 1 ampere of current for a period of 8 hours, the battery energy rating is given as E = 1 ampere × 8 hours = 8 **ampere-hours**. To convert this to joules, you first have to multiply by the battery's voltage to convert the energy to watt-hours. (Remember that volts × amperes = watts.) Assume, for example, a 1.5-volt battery rated at 8 ampere-hours. The total energy in watt-hours is calculated as follows:

$$E = 8 \text{ ampere-hours} \times 1.5 \text{ volts} = 12 \text{ watt-hours}$$

Then, to convert it to joules, you multiply by the number of seconds in an hour:

$$E = 12 \text{ watt-hours} \times 3,600 \text{ seconds/hour} = 43,200 \text{ watt-seconds}$$

Since 1 watt-second is equal to to 1 joule and since 1,000 watts = 1 kilowatt, then the total energy stored in the battery is given as

$$E = 43,200 \text{ watt-seconds} \times \frac{1 \text{ joule}}{\text{watt-second}} \times \frac{1 \text{ kilowatt}}{1,000 \text{ watts}} = 43.2 \text{ kilojoules}$$

Table 9–1 lists the conversion factors for various units of energy and power.

Table 9–1   Conversion Factors for Various Units of Energy and Power

1 horsepower	746 watts
1 horsepower	550 ft-lb/sec
1 BTU/hr × 0.293	watts
1 cal/sec	4.19 watts
1 ft-lb/sec	1.36 watts
1 BTU	1050 joules
1 joule	0.2389 cal
1 cal	4.186 joules
1 watt	0.00134 horsepower
1 watt	3.1412 BTU/hr
1 watt/sec	1 joule

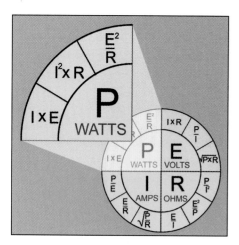

FIGURE 9–1 Ohm's law pie chart.

9.3 The Ohm's Law Pie Chart

The power delivered to a DC series circuit is a measure of the amount of work done by that circuit per unit of time. Some energy is delivered to the load and performs useful work, while other energy delivered to the circuit is dissipated as heat and is wasted in the circuit (unless, of course, the circuit is designed to generate heat). In electrical systems, you will almost always be concerned with the rate at which the energy is being developed, that is, power.

The amount of lost energy is an important consideration when designing a circuit. Remember that lost energy is energy not actually used to perform work for which the circuit was designed. In this lesson, you will examine circuits in which there are energy losses, and you will learn to calculate those power values associated with those energy losses.

Remember from previous chapters that you used the pie chart in Figure 9–1 to obtain the equations needed to calculate the different parameters in a DC circuit. You are now going to use the power slice. You can see that to calculate power, you need only two of the three parameters you have studied so far. The equations to calculate power are

$$P = I^2R$$

$$P = EI$$

$$P = \frac{E^2}{R}$$

$P = EI$ is the formula that you will use in most of your electrical work because the voltage and current are either known and/or more easily measured than the resistance of power system loads.

■ POWER IN A DC CIRCUIT

9.4 Power Rating

We have mentioned a couple of times that some of the energy in a circuit is wasted. This is the energy that goes into generating heat in the components of the circuit. For this reason, circuit components have power ratings. This allows designers and technicians to use components that will withstand the heat generated in the circuit without being damaged or significantly changing values.

EXAMPLE

The resistor in Figure 9–2 has a rating of 3 watts. Would the circuit power damage it?

FIGURE 9–2   Calculating power in a simple circuit.

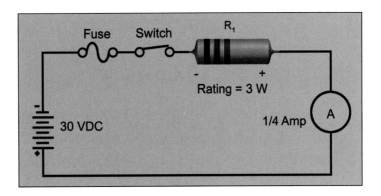

Solution:

To determine if the circuit power requirement exceeds the rating of the resistor, calculate the power being generated. Since we know current and voltage, the equation to use from Figure 9–1 is

$$P = EI$$

$$P = 30 \times 0.25$$

$$P = 7.5 \text{ watts}$$

This example shows that the circuit would damage the resistor. Another resistor of higher rating is needed for this circuit. Notice that this example ignores any losses that may occur in the wiring. However, such losses will usually be small and often can be ignored.

9.5 Circuit Calculations

Look at Figure 9–3. This circuit has multiple components with varying resistances. They range from 0 Ω for the fuse and switch to 1,500 Ω for R_4.

FIGURE 9–3   Calculating power in a more complex circuit.

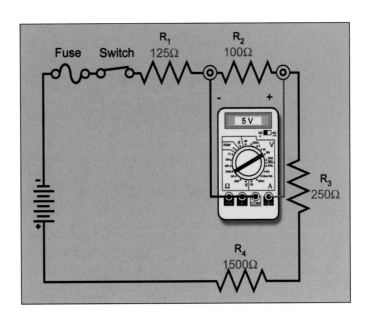

Getting Started

To determine the power requirements of all the circuit components, two parameters have to be known about each. In Figure 9–3, only one component has two parameters associated with it: R_2. Remember that in a series circuit, the current flowing through any component is the same value flowing through all the components.

Since you know the voltage drop across R_2 and you know R_2's resistance value, R_2's current is

$$I_2 = \frac{E_2}{R_2} = \frac{5}{100} = 0.05 \text{ amperes}$$

Since this is a series circuit, you know that all currents are equal, so $I_T = 0.05$ amperes.

Power Determination

The power requirement for the whole circuit can be determined by adding the power for each individual component. Start first with the fuse and switch. The total current through the circuit and the fuse and switch are assumed to have 0 Ω resistance each; then, from the pie chart in Figure 9–2,

$$P = I^2R$$
$$P = (0.05)^2 \,(0) = 0 \text{ watts}$$

This shows that without resistance, a component will not dissipate any real power. Although there is current flowing through it, the component must have resistance to dissipate power. The power consumed by each resistor is

$$P_{R_1} = I_{R_1}^2 \times R_1 = (0.05)^2 \times (125) = 0.3125 \text{ W}$$
$$P_{R_2} = I_{R_2}^2 \times R_2 = (0.05)^2 \times (100) = 0.25 \text{ W}$$
$$P_{R_3} = I_{R_3}^2 \times R_3 = (0.05)^2 \times (250) = 0.625 \text{ W}$$
$$P_{R_4} = I_{R_4}^2 \times R_4 = (0.05)^2 \times (1,500) = 3.75 \text{ W}$$

Next determine the voltage of the battery. You can do this in a couple of different ways. First you know that the total voltage is equal to the sum of all the voltage drops in a series DC circuit (chapter 7). Going back to the pie chart, voltage can be calculated in one of three ways. Since you know the resistance of each resistor as well as the current through each resistor, the easiest calculation is to multiply the current times the resistance. For each resistor, this gives

$$E_{R_1} = I_{R_1} \times R_1 = 0.05 \times 125 = 6.25 \text{ V}$$
$$E_{R_2} = I_{R_2} \times R_2 = 0.05 \times 100 = 5 \text{ V}$$
$$E_{R_3} = I_{R_3} \times R_3 = 0.05 \times 250 = 12.5 \text{ V}$$
$$E_{R_4} = I_{R_4} \times R_4 = 0.05 \times 1,500 = 75 \text{ V}$$

The total voltage of the circuit is equal to the battery voltage and can be calculated by adding the four resistor voltages:

$$E_T = E_1 + E_2 + E_3 + E_4 = 6.25 + 5 + 12.5 + 75$$
$$E_T = 98.75 \text{ V}$$

You also can calculate the total voltage by using the power formulas. Like V_T and R_T in a DC circuit, the power consumed by the entire circuit is the sum of the power used by each component. So,

$$P_T = P_1 + P_2 + P_3 + P_4 = P_T = 0.3125 + 0.25 + 0.625 + 3.75$$
$$P_T = 4.9375 \text{ W}$$

Therefore,

$$E = \frac{P}{I} = \frac{4.9375}{0.05} = 98.75 \text{ V}$$

■ SUMMARY

The unit of electrical power is the watt. The definition of electrical power adopted by the United States is based on the international standard of 1 joule per second. When 1 joule of work is done in 1 second, 1 watt of power is used. Power is the rate of doing work.

In an electrical circuit, the electrical power used can be calculated by using the formula

Power (watts) = current (amperes) × voltage or EMF (volts)

which is represented as

$$P = I \times E$$

The ampere is the SI (International System of Units) unit for current and is defined as

1 ampere = 1 coulomb/second

The volt is the recognized unit for electrical potential and is defined as

1 volt = 1 joule/coulomb

Given these relationships, we determined that

Power (watt) = coulomb/second × joule/coulomb = joule/second

The SI unit for electrical power is the watt (W).

If Ohm's law is used to substitute for specific units in the power formula, two other equations for electrical power can be derived:

Given

$$P = I \times E$$

and substituting for I

$$I = \frac{E}{R}$$

then

$$P = \frac{E}{R} \times E = \frac{E^2}{R}$$

and substituting for E

$$E = I \times R$$

then

$$P = I \times IR = I^2R$$

Depending on the parameters known about a circuit, the use of any of these power formulas will result in a correct answer. At the same time, if the power rating and the resistance value of a resistor are known, the maximum value of current that can flow through the resistor can be determined as well as the maximum allowable voltage drop:

Maximum current:

$$I = \sqrt{\frac{P}{R}}$$

Maximum voltage drop:

$$E = \sqrt{PR}$$

■ REVIEW QUESTIONS

1. Power is measured in ____.
2. A watt is defined as ____.
3. What is (are) the formula(s) for finding watts in a DC series circuit?
4. What power formula will you use out in the field most of the time?
5. Is it necessary to have a resistor rated for the circuit wattage? If so, why?

6. One horsepower equals how many watts?
7. If you know the power dissipation for each element in a series circuit, how can you calculate the total power dissipation in the whole circuit?
8. Discuss the unit of horsepower and where it originated.

■ PRACTICE PROBLEMS

1. The following questions are based on a ½-horsepower motor rated at 120 volts.
   a. How many amperes will the motor draw?
   b. What size resistor (in ohms) would draw the same amount of power?
   c. Draw the equivalent circuit for the motor.
2. Consider the following circuit and calculate all the unknown values:

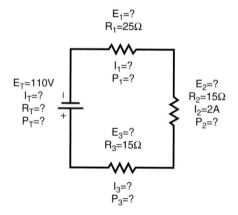

3. A certain commercially made 480-V UPS battery is rated at 1,000 ampere-hours. How many kilojoules of energy are stored in this battery?

PART

3

DC PARALLEL CIRCUITS

chapter **10**

How Voltage Functions in a DC Parallel Circuit

■ **OUTLINE**

OVERVIEW

In all the previous chapters, you dealt only with series DC circuits. This helped you develop an understanding of Ohm's law and all the associated characteristics and equations. Using the tools you now have for determining voltage, resistance, current, and power, you can move into a new area of electricity: parallel circuits.

This chapter deals with how voltage is applied in parallel circuits. One of the best examples of parallel circuits is house wiring. If a house were wired in series, all the appliances, lights, and receptacles would have to be turned on at one time for any one of them to work. If an iron overheated and burned out, all the rest of the appliances and lights would go out. Houses are wired with parallel circuits—circuits that have more than one path for current. In this chapter you will see how the applied voltage is impressed across parallel loads and how the voltage for each load is calculated.

OBJECTIVES

After completing this chapter, you should be able to:

1. Draw and identify components in parallel circuits.
2. Identify and describe differences between voltage sources in series and parallel circuits.
3. Solve problems involving voltage in parallel circuits using Ohm's law.

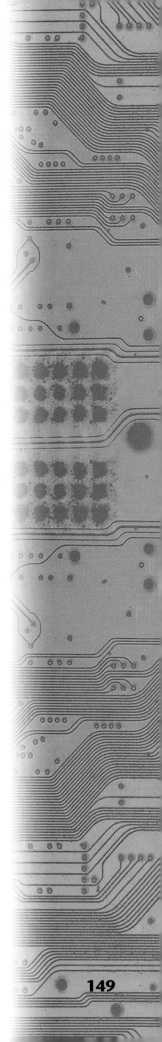

■ PARALLEL CIRCUITS

10.1 Definition and Principles

Parallel circuits have more than one path for current. These paths are called *branches*. All the parallel branches have the same voltage applied, and each branch contributes to the total circuit current. The total current is made up of the sum of the current flows from each branch of the circuit (see Figure 10–1).

The colored arrows indicate the direction of current flow. You can see the current takes two paths to move from the negative side of the battery back to the positive side. The circuit is called *parallel* because the branches are connected side by side, or parallel to each other. Both ends of each branch are connected to each side of the battery.

Figure 10–2 is the same circuit without the current arrows. Notice that the two resistors have voltmeters connected across them. The voltmeters are measuring the voltage drop across each branch. Each voltmeter relates to the battery in the same way. The top probe of each voltmeter is connected through the switch and fuse to the negative side of the battery. The bottom probe is connected to the positive side of the battery.

With both voltmeters connected to the same place on the battery, each reads the same voltage. In other words, both voltmeters are reading the battery voltage. You can see that the voltmeters are also reading the voltage drop across each resistor. This means that the parallel branches have the same voltage applied to them: battery voltage.

FIGURE 10–1   A simple parallel circuit.

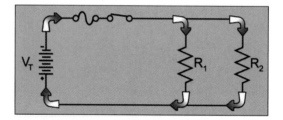

FIGURE 10–2   Voltage readings across parallel resistors.

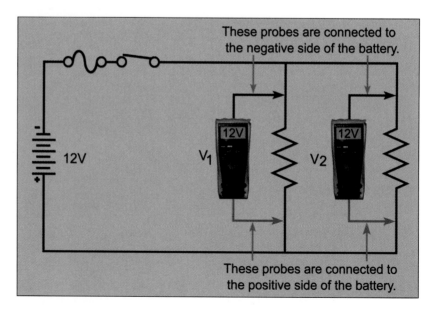

The basic fact about parallel circuit branches is that they all have the same applied voltage. This is true regardless of the number of branches connected in parallel. As long as they are all connected to the same voltage source, they will have the same applied voltage:

$$V_1 = V_2 = V_3 = \ldots V_n$$

where n is the number of branches in the parallel circuit.

Figure 10–3 shows a simple parallel house circuit as an example of how parallel circuits work. Note that the toaster and the light are connected in parallel with the house voltage supply using the wall receptacle.

10.2 Multiple Voltage Sources

Circuit branches are not the only components that may be connected in parallel. Voltage sources may also be connected this way. One of the most common reasons for connecting voltage sources in parallel is to supply more current to a circuit.

All *real* power supplies have power limits. Both generators and batteries have maximum amounts of current that they can supply to a load. If a circuit requires higher current but the same voltage, then sources can be connected in parallel to combine their currents. Generators are connected in parallel all across a power system for this very reason. Figure 10–4 shows how batteries can be connected in parallel. Notice that each battery will supply its rated current to the lights; consequently, the total energy capability for the lights is tripled.

Figure 10–5 is another example of voltage sources connected in parallel. Each voltage source in the circuit is a 12-volt battery rated at 80 amp-hours. All the negative ends are connected together, and all the positive ends are connected together. The result is a source voltage of 12 VDC with a combined load capacity of 240 amp-hours.

FIGURE 10–3   Simple house power parallel circuit.

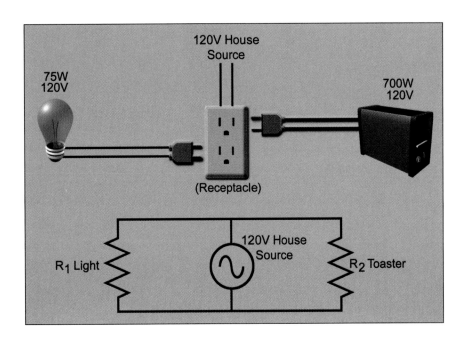

FIGURE 10–4   Batteries in parallel.

FIGURE 10–5   Voltage sources connected in parallel.

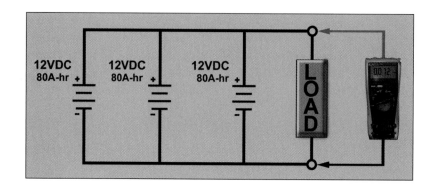

The voltmeter is measuring all the voltage sources. Since all the terminals of the batteries are connected, the voltmeter senses one voltage source in the circuit. In this circuit, the voltmeter reads 12 VDC. The voltage output from the voltage sources is equal to any individual battery. This would be true regardless of the number of batteries connected in the circuit.

CAUTION: Connecting voltage sources in parallel that do not have the same voltages and polarities is an unsafe practice and should never be done. Excessive currents could be produced that would damage the voltage sources and cause personal injury and explosions.

10.3 How Voltage Relates to Current through Parallel Branches

This chapter deals mainly with how voltage functions in parallel circuits. However, as a preview to coming chapters, it is fitting here to discuss current flow through parallel DC circuits. Earlier the fact was discussed that parallel paths in the circuit share (or divide) the total current in a parallel circuit. The amount of current is dependent on the amount of resistance in each particular path or branch (i.e., the higher the resistance of the path, the lower the current). The total current is simply the sum of all the parallel branch currents. Current flow in parallel DC circuits will be discussed in detail in chapter 12.

Figure 10–6 shows a simple parallel circuit with two branches. The first branch has 0.5 amps of current flowing through it. Knowing this value allows you to determine the voltage drop across resistor R_1. If the voltage across R_1 is known, then the voltage across R_2 is known since R_1 and R_2 are in parallel. Finally, because the battery, R_1, and R_2 are all in parallel, calculating the voltage across R_1 allows the determination of the source voltage.

FIGURE 10–6   Parallel circuit
with current shown in one branch.

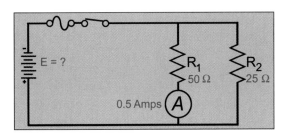

EXAMPLE

What is the voltage drop across R_1 and R_2? What is the supplied voltage?

Solution:
First, use Ohm's law to determine the voltage across R_1:

$$E_1 = I_{R_1} \times R_1 = 0.5 \times 50 = 25 \text{ V}$$

Then, since R_1 and R_2 are in parallel,

$$E_1 = E_2 = 25 \text{ V}$$

And finally, since R_1, R_2, and the battery are all in parallel,

$$E_1 = E_2 = \text{battery voltage} = 25 \text{ volts}$$

■ SUMMARY

Parallel circuits have more than one path for current to flow. These different paths are called parallel branches. The voltage applied to the parallel branches is equal to the voltage source. The current through each branch is less than the total current. The sum of the branch currents is equal to the total current.

$$I_T = I_1 + I_2 + I_3 + \ldots I_n$$

Voltage sources may also be connected in parallel as long as they are the same value and of the same polarity. When a circuit has more than one voltage source, the output voltage from the parallel sources is equal to the voltage from any one of the sources:

$$V_{\text{battery}} = V_1 = V_2 = V_3 = \ldots V_n$$

■ REVIEW QUESTIONS

1. Parallel circuits provide more than one path for current to flow. Discuss how the current flows in each parallel branch.

2. What is the formula for finding the total voltage in a series circuit? A parallel circuit?

3. What is the formula for finding the current in a series circuit? A parallel circuit?

4. The voltage across each branch of a parallel circuit is the same. Why?

■ PRACTICE PROBLEMS

1. A parallel circuit contains three resistors. R_1 has 10 volts dropped across it, R_2 has 10 volts dropped across it, and R_3 has 10 volts dropped across it.

   a. What is the total voltage supplied to the circuit?

   b. Draw the circuit.

2. A parallel circuit contains two resistance loads. Load 1 is rated 100 ohms, and load 2 is rated 50 ohms and draws 2 amps.

   a. What is the voltage across each branch?

   b. What is the total current through each branch?

   c. What is the total voltage?

   d. What is the total current?

   e. What is the total resistance?

   f. Draw the circuit.

chapter 11

Understanding Resistance in a DC Parallel Circuit

■ **OUTLINE**

■ OVERVIEW

Chapter 10 introduced you to parallel circuits and how voltage is applied across the branches in the circuit. In this chapter, you will learn different methods of calculating resistance in parallel circuits. You will recall that to calculate the total resistance in series, you simply added the values of all the resistors:

$$R_T = R_1 + R_2 + R_3 + \ldots R_n$$

From this equation, it seems sensible to think that whenever resistance is added to a circuit, total resistance will increase and total current will decrease. This is true only for a series circuit. We will show that in a parallel circuit, as resistors are added, the total resistance decreases and total current increases.

In chapter 10, you learned that a parallel circuit is one in which there is more than one path for current. These paths are called *branches.* Each branch in a parallel circuit has the same voltage drop (applied voltage) across it. Each branch does not have to have the same resistance. This means that the current flowing through each branch depends on the resistance of that branch. As stated in chapter 10, the total current in the circuit is simply the sum of the various currents flowing through each of the branches.

■ OBJECTIVES

After completing this chapter, you should be able to:

1. Draw circuits containing parallel resistors.
2. Calculate the total circuit resistance of parallel circuits with two resistance values using the product ÷ sum method.
3. Calculate the total circuit resistance of parallel circuits with two or more resistance values using the reciprocal method.

■ GLOSSARY

Product over sum equation An equation that can be used to calculate the equivalent resistance of two resistors in parallel. It is a simplification of the reciprocal equation: $R_T = \dfrac{R_1 \times R_2}{R_1 + R_2}$

Reciprocal equation An equation that can be used to calculate the equivalent resistance of resistors in parallel:

$$\frac{1}{R_T} = \frac{1}{R_1} + \frac{1}{R_2} + \frac{1}{R_3} + \cdots \frac{1}{R_n}$$

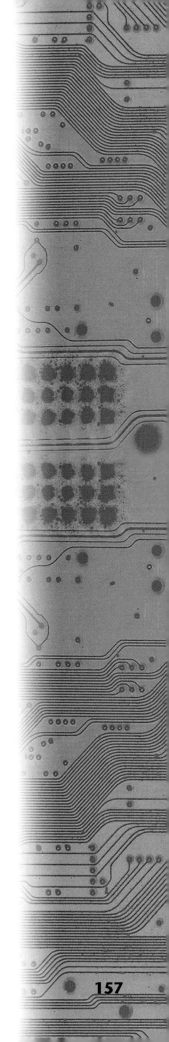

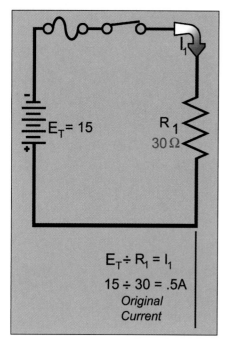

$$E_T \div R_1 = I_1$$
$$15 \div 30 = .5A$$
Original
Current

FIGURE 11–1   Single-resistor circuit.

■ CALCULATING TOTAL RESISTANCE IN PARALLEL CIRCUITS

11.1 Parallel Resistance

The fact that total circuit resistance decreases as resistance is added to a parallel circuit is easily misunderstood. Figures 11–1 to 11–3 show a circuit with a progression of parallel resistors being added to the circuit. Note that each branch has the same resistance and thus the same current. Also note the effect that adding each resistor has on the total current of the circuit.

In Figure 11–1, you see a relatively simple circuit comprising a resistor in series with a fuse and switch. You can see that the source voltage is the same voltage that is seen by resistor R_1. Therefore, $E_T = E_1$. Use Ohm's law to calculate the current through the resistor R_1:

$$E_1 = I_1 \times R_1$$

$$I_1 = \frac{E_1}{R_1} = \frac{15}{30} = 0.5 \text{ A}$$

When you look at the total circuit current in Figure 11–1, you see that the only current in the circuit is passing through resistor R_1. That means that $I_1 = I_T$; therefore,

$$I_T = I_1 = 0.5 \text{ A}$$

Now look at Figure 11–2. A second resistor has been added in parallel with the resistor in the circuit from Figure 11–1. The second resistor is seeing the same voltage across it as the resistor in Figure 11–1. In parallel circuits, the voltage is the same across all branches of the circuit. In Figure 11–2, we have added a second branch, and that

FIGURE 11–2   Adding a second resistor in parallel.

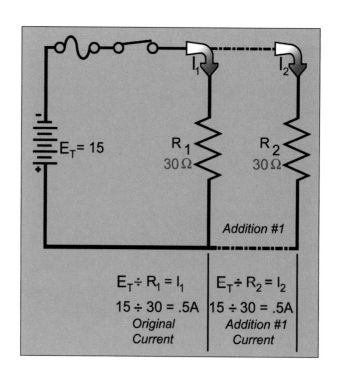

$$E_T \div R_1 = I_1 \qquad E_T \div R_2 = I_2$$
$$15 \div 30 = .5A \quad 15 \div 30 = .5A$$
Original   *Addition #1*
Current   *Current*

FIGURE 11–3 Multiple resistors in parallel.

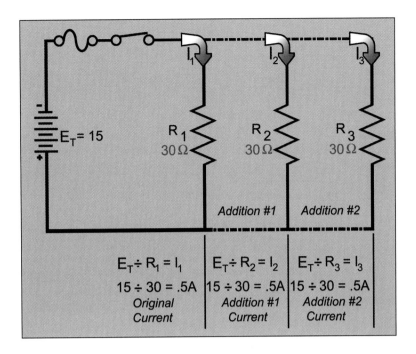

branch is also seeing the supply voltage E_T. Therefore, $E_2 = E_T = E_1$. To calculate the current through the resistor R_2, you again use Ohm's law:

$$I_2 = \frac{E_2}{R_2} = \frac{15}{30} = 0.5 \text{ A}$$

By adding the second resistor, the total circuit current changed. Remember,

$$I_T = I_1 + I_2 + \ldots I_n$$

Therefore,

$$I_T = I_1 + I_2 = 0.5 + 0.5 = 1 \text{ A}$$

By adding a second resistor in parallel with the first resistor, you have also added to the total current of the circuit.

Now add a third resistor. Figure 11–3 shows the new circuit with three resistors in parallel. Just as both the first and the second resistors saw the supply voltage E_T across them, the third resistor will also see the supply voltage across it. That is one of the basic rules of parallel circuits: All branches in a parallel circuit will see the same voltage across each of the branches. The voltage that they will see is the voltage supplying that branch.

You now calculate the current in the third branch, again using Ohm's law:

$$I_3 = \frac{E_3}{R_3} = \frac{15}{30} = 0.5 \text{ A}$$

With the third resistor, the total circuit current has changed. Remember,

$$I_T = I_1 + I_2 + I_3 \ldots I_n$$

Therefore,

$$I_T = I_1 + I_2 + I_3 = 0.5 + 0.5 + 0.5 = 1.5 \text{ A}$$

By adding a third resistor in parallel with the first and second resistors, the total current of the circuit has increased. In Figure 11–3, it just so happens that all the resistances were equal, but not all circuits are so convenient. As resistors are added in parallel, regardless of their value the total circuit current increases. The current from each branch will make up a part of the total of all the branches.

Using Ohm's law, the total resistance of all the branches of a parallel circuit can also be calculated once the total circuit current is known. The following calculations show how to find the total circuit resistance. Note that the total circuit resistance value is smaller than any of the individual branch resistance values.

Ohm's law states that

$$R_T = \frac{E_T}{I_T} \text{ and } I_1 + I_2 + I_3$$

Therefore,

$$R_T = \frac{15 \text{ V}}{0.5 \text{ A} + 0.5 \text{ A} + 0.5 \text{ A}} = \frac{15 \text{ V}}{1.5 \text{ A}} = 10 \ \Omega$$

The piping system shown in Figure 11–4 can be compared to a parallel electrical circuit. The pump provides the force for flow like a battery. The pipes in parallel are like the parallel branches. Imagine now that as the parallel pipes are opened to the main pipe, the "hole" for fluid to flow through gets larger. The restriction to flow for the fluid goes down and the pump can do more work. This is the same for parallel circuits. As more branches are added, the total restriction to current flow goes down, and the total current flow goes up.

FIGURE 11–4   Parallel piping.

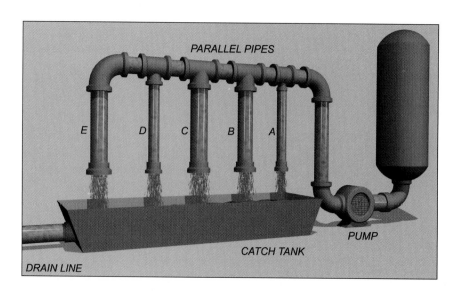

11.2 Parallel Resistance Reciprocal Equation

To calculate the total resistance in a parallel circuit, a general formula is needed. This formula is known as the reciprocal equation. This equation may be used for any parallel circuit. The following describes how the reciprocal equation is derived. In Figure 11–5, the sum of all the branch currents is equal to the total current:

$$I_T = I_1 + I_2 + I_3$$

Using Ohm's law and substituting the equation for current in the previous equation,

$$\frac{E_T}{R_T} = \frac{E_1}{R_1} + \frac{E_2}{R_2} + \frac{E_3}{R_3}$$

Since this circuit is a parallel circuit, all the branch voltages are equal to each other and to the total voltage, so

$$\frac{E_T}{R_T} = \frac{E_T}{R_1} + \frac{E_T}{R_2} + \frac{E_T}{R_3}$$

If you divide both sides of this equation by E_T, the equation becomes

$$\frac{1}{R_T} = \frac{1}{R_1} + \frac{1}{R_2} + \frac{1}{R_3} \tag{1}$$

This is the reciprocal equation, but it results in the reciprocal of total resistance. To get total resistance, take the reciprocal of both sides. Equation 1 then becomes

$$R_T = \frac{1}{\dfrac{1}{R_1} + \dfrac{1}{R_2} + \dfrac{1}{R_3}}$$

The general form of Equation 1 is

$$\frac{1}{R_T} = \frac{1}{R_1} + \frac{1}{R_2} + \frac{1}{R_3} \cdots \frac{1}{R_n} \tag{2}$$

where n is the number of branches in the circuit.

FIGURE 11–5   Current flow in multiple parallel resistors.

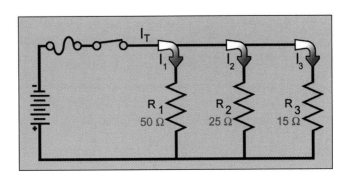

EXAMPLE 1

What is the total resistance in Figure 11–5?

Solution:
To calculate the total resistance, use the reciprocal equation (Equation 1):

$$\frac{1}{R_T} = \frac{1}{R_1} + \frac{1}{R_2} + \frac{1}{R_3} = \frac{1}{50} + \frac{1}{25} + \frac{1}{15} = 0.1267$$

Therefore,

$$R_T = \frac{1}{0.1267} = 7.89 \ \Omega$$

Notice that the total resistance (R_T) is less than the smallest branch resistance (15 Ω).

11.3 Product over Sum Equation

If the parallel circuit has only two branches, a more simple equation can be used. Look at the reciprocal equation again. If the circuit has only two branches, the equation becomes

$$R_T = \frac{1}{\dfrac{1}{R_1} + \dfrac{1}{R_2}}$$

At this point it is useful to find a common denominator for the right-hand side of the equation. This becomes

$$R_T = \frac{1}{\dfrac{R_2}{R_1 R_2} + \dfrac{R_1}{R_1 R_2}} = \frac{1}{\dfrac{R_1 + R_2}{R_1 R_2}} = \frac{R_1 R_2}{R_1 + R_2}$$

$$R_T = \frac{R_1 R_2}{R_1 + R_2} \tag{3}$$

Equation 3 is called the **product over sum equation** and can be used for any circuit in which there are only two branches.

EXAMPLE 2

What is the total resistance in Figure 11–6?

$$R_T = \frac{R_1 R_2}{R_1 + R_2} = \frac{40 \times 30}{40 + 30} = \frac{1,200}{70} = 17.14 \ \Omega$$

11.4 Resistors with Equal Value Equation

Parallel circuits with resistors of equal value are the easiest to calculate. The total resistance is simply the value of one of the resistors divided by the number of branches in the circuit. Refer to Figure 11–7.

FIGURE 11–6 Two parallel branches.

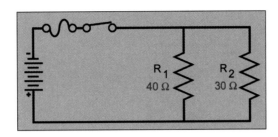

FIGURE 11–7 Parallel branches with equal resistances.

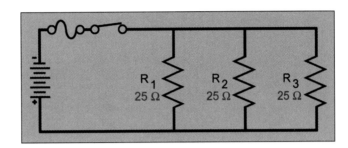

Use the reciprocal equation to determine how to calculate the total resistance of a parallel circuit whose branches are all of equal resistance:

$$R_T = \cfrac{1}{\cfrac{1}{R_1} + \cfrac{1}{R_2} + \cfrac{1}{R_3}}$$

Since R_1, R_2, and R_3 are equal, the equation becomes the following after substituting R_1 for the other resistance values:

$$R_T = \cfrac{1}{\cfrac{1}{R_1} + \cfrac{1}{R_1} + \cfrac{1}{R_1}} = \cfrac{1}{\cfrac{3}{R_1}} = \cfrac{R_1}{3}$$

The general form of this equation is

$$R_T = \frac{R_X}{n}$$

where R_X is the resistance of any branch and n is the number of branches.

EXAMPLE 3

What is the total resistance of the circuit in Figure 10–7?

$$R_T = \frac{R_X}{n} = \frac{25}{3} = 8.33 \ \Omega$$

Prove the calculation using the reciprocal method:

$$R_T = \cfrac{1}{\cfrac{1}{R_1} + \cfrac{1}{R_2} + \cfrac{1}{R_3}} = \cfrac{1}{\cfrac{1}{25} + \cfrac{1}{25} + \cfrac{1}{25}} = \frac{25}{3} = 8.33 \ \Omega$$

FIGURE 11–8 Resistors in parallel that are multiples of 36 Ω.

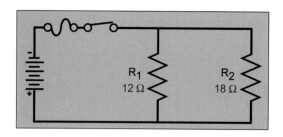

FIGURE 11–9 Equivalent circuit of Figure 11–8.

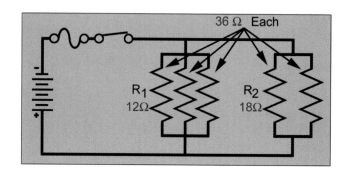

11.5 Equal Value Variation

As you have studied, the equal resistance rule applies when all the resistance values in a parallel circuit are the same. A variation of the equal resistance rule can also be used when there are different resistance values in a parallel circuit and when these resistance values are multiples of one another or can be divided evenly into a common multiple.

For example, any single resistor can be considered as two or more equal value resistors connected in parallel. A single 12-Ω resistor can be thought of as two 24-Ω resistors or three 36-Ω resistors connected in parallel, and an 18-Ω resistor could be thought of as two 36-Ω resistors or three 54-Ω resistors.

If you have a 12-Ω resistor and an 18-Ω resistor connected in parallel, you can express the circuit as three 36-Ω resistors in parallel (12 Ω = 36 Ω ÷ 3) with two more 36-Ω resistors (18 Ω = 36 Ω ÷ 2) in parallel. Refer to Figure 11–8.

This circuit can also be shown as equal value resistors in parallel. Figure 11–9 shows the equivalent circuit using a common dividend (36) as the value of resistors in parallel. The total resistance can now be calculated using the equal resistance equation:

$$R_T = \frac{R_X}{n} = \frac{36}{5} = 7.2 \ \Omega$$

■ SUMMARY

The total resistance for resistors in parallel is always less than the smallest resistor. Adding more resistors to branches in a parallel circuit causes the total resistance to decrease and the total current to increase.

Total resistance of a parallel circuit can be calculated in different ways. For any parallel circuit, the reciprocal equation is

$$R_T = \cfrac{1}{\cfrac{1}{R_1} + \cfrac{1}{R_2} + \cfrac{1}{R_3}}$$

For a two-branch circuit, the product over sum equation is

$$R_T = \frac{R_1 R_2}{R_1 + R_2}$$

And for circuits with equal values, the equation is

$$R_T = \frac{R_X}{n}$$

■ REVIEW QUESTIONS

1. In a DC parallel circuit, if resistance is added, what happens to the total resistance?
2. If resistance is added in a DC parallel circuit, what happens to the total current?
3. What is the reciprocal formula for finding total resistance in a parallel DC circuit?
4. What is the formula for finding total resistance in a DC series circuit?
5. What is the product over sum equation for finding total resistance in a DC parallel circuit?
6. What is the equation for finding total resistance in a DC parallel circuit when all resistive loads are the same?
7. What is the relationship between the total resistance in a parallel circuit and the value of the smallest resistor in the parallel combination?

■ PRACTICE PROBLEMS

1. A DC circuit has the resistances hooked in parallel. One branch is rated at 20 ohms, one branch at 40 ohms, and one branch at 60 ohms.
   a. Draw the circuit.
   b. Find the total resistance using the reciprocal method.
2. A DC parallel circuit has two loads. Load 1 is rated at 10 Ω, and load 2 is rated at 5 Ω. The circuit is fed with 24 volts.
   a. Draw the circuit.
   b. Calculate the total resistance of the parallel circuit.
   c. Calculate the current flow through each load.
   d. Calculate the voltage across each load.
3. A DC parallel circuit has three loads. Branch 1 is 360 Ω, branch 2 is 240 Ω, and branch 3 is 144 Ω. The circuit has a 120-volt supply.
   a. Draw the circuit.
   b. Label each branch with its current.
   c. Calculate the total resistance.

How Current Reacts in a DC Parallel Circuit

■ OUTLINE

■ OVERVIEW

This chapter will extend your knowledge of DC parallel circuits by showing you how current flows through multiple branches. You already have the tool required to analyze current flow in a parallel circuit—Ohm's law. Now you will learn how to combine Ohm's law analysis with the various total resistance theorems from chapter 11 and calculate all the current flow in a parallel circuit.

■ OBJECTIVES

After completing this chapter, you should be able to:

1. Draw parallel circuits showing alternate current paths in those circuits.
2. Calculate the currents in individual branches of parallel circuits.
3. Determine the total current in parallel circuits by determining the equivalent resistance of the circuit and solving for current.
4. Determine total current by finding the current in each branch and then solving for total current.

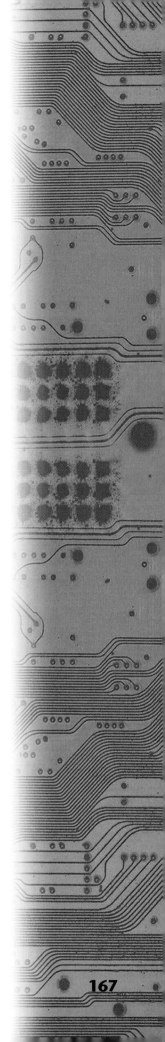

■ THEORY AND CALCULATION OF CURRENT FLOW IN PARALLEL CIRCUIT BRANCHES

12.1 Parallel Resistance Review

Your previous studies in parallel circuits introduced you to voltage and resistance. You have calculated how the voltage is impressed across the branches and across the components in the branch. The branch voltage is always equal to the total voltage and to all the other branch voltages:

$$E_T = E_1 = E_2 = E_3 = \ldots E_n$$

You calculated the parallel resistance and total resistance using three equations:

The reciprocal equation:

$$R_T = \cfrac{1}{\cfrac{1}{R_1} + \cfrac{1}{R_2} + \cfrac{1}{R_3} \cdots \cfrac{1}{R_n}}$$

The product over sum equation:

$$R_T = \frac{R_1 R_2}{R_1 + R_2}$$

The equal value equation:

$$R_T = \frac{R_x}{n}$$

You know that the total current in a parallel circuit is equal to the sum of the branch currents. In this chapter, you will explore current in depth and learn how to calculate current for all the components in a parallel DC circuit using just a few known values. You have learned all the equations you will need to perform these calculations. Recall the Ohm's law pie chart as shown in Figure 12–1. This chart is a reference for all the variations of calculating voltage, current, resistance, and power. In this chapter focus on the current calculations.

12.2 Theory of Current in Parallel

As described previously, current flow in parallel can be thought of as resembling flow through a piping system. Figure 12–2 represents a parallel piping system. The arrows show the relative amount of flow through the pipes. As you move from right to left (A to E) in the upper main pipe, the amount of flow in that pipe decreases because some of the flow branches off into each parallel branch. Moving left to right (E to A) in the lower main pipe shows the flow increasing. Everything that leaves the pump goes through the system of pipes and reenters the pump. This means that both the upper and the lower main pipes have the same amount of flow.

The path between the two main pipes is the combination of the five parallel pipes (labeled A through E). Think of the main pipe as spreading out into five smaller pipes and then recombining back into one

FIGURE 12-1   Ohm's law pie chart

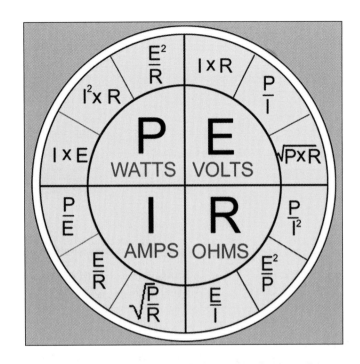

FIGURE 12-2   A parallel piping system

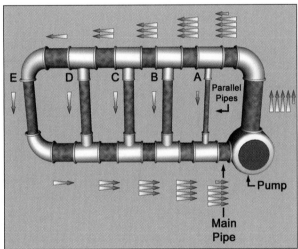

pipe before entering the pump. The combined flow through the parallel pipes is equal to the flow through the main pipes. In other words, the sum of the flow of the parallel pipes = total flow:

Flow in Pipe A + Pipe B + Pipe C + Pipe D + Pipe E = Total Flow

The same is true about parallel DC circuits. The total current flow leaving the voltage source is equal to the combination of the currents in all the branches:

$$I_T = I_1 + I_2 + I_3 + \ldots + I_n \qquad (1)$$

where n is the number of branches in the parallel circuit.

In Figure 12–3, the total current (I_T) is shown leaving the battery and returning to the battery. Between the two points where current equals I_T, the current divides into three parallel circuit branches, each containing a resistor (R_1, R_2, and R_3). After traveling through each branch, the current recombines into I_T before reentering the battery.

FIGURE 12–3   Current flow in a
parallel circuit

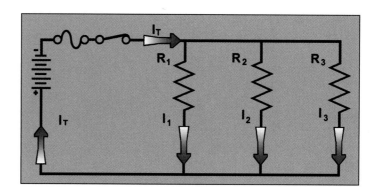

FIGURE 12–3   Current flow in a
parallel circuit

12.3 Calculating Current in Parallel

The remainder of this chapter provides examples of parallel circuits
in which current will have to be calculated. Figure 12–4 is the first
example.

EXAMPLE 1

What is the current through each of the three branches in Figure 12–4?

Solution:
To calculate the current, determine what the known values are:

$R_1 = 15 \ \Omega, R_2 = 30 \ \Omega, R_3 = 45 \ \Omega$

$E = 25$ VDC

Voltage across each branch $= 25$ VDC

From Ohm's law,

$$I_1 = \frac{E_T}{R_1} = \frac{25}{15} = 1.667 \text{ A}$$

$$I_2 = \frac{E_T}{R_2} = \frac{25}{30} = 0.833 \text{ A}$$

$$I_3 = \frac{E_T}{R_3} = \frac{25}{45} = 0.556 \text{ A}$$

FIGURE 12–4   Calculating
current in a parallel circuit

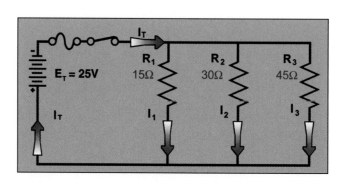

EXAMPLE 2

What is the total current in Figure 12–4?

Solution:
There are two methods for determining total current. The first uses the reciprocal equation. Replacing branch resistance with total resistance will yield total current. The first step is to calculate total resistance:

$$R_T = \cfrac{1}{\cfrac{1}{R_1} + \cfrac{1}{R_2} + \cfrac{1}{R_3}} = \cfrac{1}{\cfrac{1}{15} + \cfrac{1}{30} + \cfrac{1}{45}} = \cfrac{1}{0.067 + 0.033 + 0.022} = 8.18\ \Omega$$

Then, using the equation for current from Ohm's law and placing total resistance in for R, the total current is

$$I_T = \frac{E_T}{R_T} = \frac{25}{8.18} = 3.056\ A$$

Another method for determining total current in Figure 12–4 is to add all the branch currents:

$$I_T = I_1 + I_2 + I_3$$
$$I_T = 1.667 + 0.833 + 0.556 = 3.056\ A$$

EXAMPLE 3

Calculate the branch currents and total current as illustrated in Figure 12–5.

Solution:
Using Ohm's law and the example from Figure 12–5, the solution is

$$I_1 = \frac{P_1}{E_T} = \frac{18}{50} = 0.36\ A$$

$$I_2 = \frac{P_2}{E_T} = \frac{11}{50} = 0.22\ A$$

$$I_T = I_1 + I_2 = 0.36 + 0.22 = 0.58\ A$$

FIGURE 12–5 Calculating power in parallel

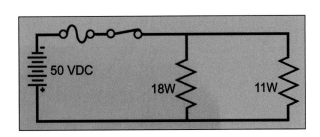

50 VDC   18W   11W

FIGURE 12–6 Power sources in parallel

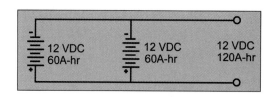

12.4 Parallel Power Sources

Power sources may also be connected in parallel. In this case, each power source provides a portion of the total circuit current. This configuration divides the current by the number of power sources connected in parallel. The advantage of placing power sources in parallel is an increased lifetime for each individual source. A single 12-V battery rated at 60 amp-hours would last twice as long if another 12-V, 60-amp-hour battery were added in parallel. Look at Figure 12–6.

With two 12-V, 60-amp-hour batteries in parallel, the total output increases to 120 amp-hours. Each battery will last twice as long in a circuit since each will be supplying half the required current.

CAUTION: Batteries of unequal voltage should never be placed in parallel. This situation would create excessive current and generate heat and could cause personal injury due to explosion.

Note that if the parallel sources are of equal power capacity (e.g., 60 amp-hours), the two will contribute equally to the total current. If they are not of equal power capacity (e.g., 30 amp-hours and 60 amp-hours), the larger one will generally contribute more of the circuit current.

■ SUMMARY

All the branches in a parallel circuit contribute to the total current. The sum of the branch circuits is equal to the total current. The total current can be calculated in several ways. If the resistance and voltage drop of each branch are known, calculate each branch current and add the results:

$$I_T = I_1 + I_2 + I_3 + \ldots I_n$$

Or you can use the method shown in Example 2, which is to calculate total resistance and use Ohm's law.

When sources (batteries) are connected in parallel, they will each contribute to the total current flow; consequently, the two together will last longer than either one separately. You should never connect sources of different voltage in parallel.

■ REVIEW QUESTIONS

1. Write and discuss the formula for finding E_T in a parallel DC circuit.

2. Write and discuss three formulas for finding total resistance in a DC parallel circuit.

3. Write and discuss the formula for finding total current in a DC parallel circuit.

4. Give the three formulas for solving current from the Ohm's law pie chart.

5. You have a car with a 60-amp-hour volt battery. If you placed a 6-volt, 30-amp-hour battery in parallel with a 12-volt battery, what would the total voltage be?

■ PRACTICE PROBLEMS

1. As shown in the following diagram, a parallel DC circuit contains three branches. Fill in the missing information in the diagram.

2. As shown in the following diagram, a parallel DC circuit contains three branches. Fill in the missing information in the diagram.

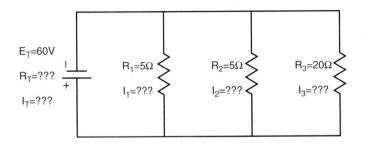

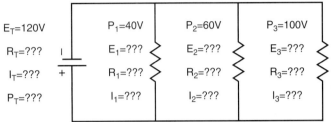

chapter 13

How Current Dividers Work in a DC Parallel Circuit

■ OUTLINE

PROPORTIONALITY   LAW OF PROPORTIONALITY
CURRENT DIVIDERS

■ OVERVIEW

E arlier you learned that voltage divides across resistors in a series circuit. The voltage division in a voltage divider circuit is proportional to the resistance of each individual resistor.

Current in parallel circuits also divides. In this chapter you will learn how to calculate the current through two branches of a current divider using simple formulas. You should practice these techniques because they will be very useful to you in your career.

■ OBJECTIVES

After completing this chapter, you should be able to:

1. Explain the difference between directly proportional and inversely proportional relationships.
2. State the Law of Proportionality as it applies to parallel circuits.
3. Solve problems involving resistors in parallel using the Law of Proportionality.

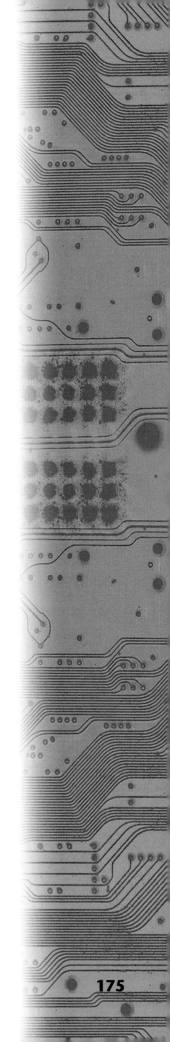

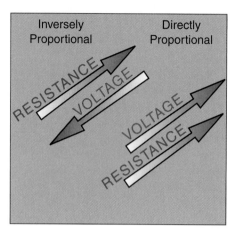

FIGURE 13–1   Direct and inverse proportionality.

■ PROPORTIONALITY

In series circuits, the voltage drop across a resistor is related to the value of the resistance. As the resistance increases, the voltage drop increases and the current through the same resistor decreases, its value changing in the opposite direction. If resistance increases, current decreases. We say that values that change in the same direction are directly proportional. Values that change in the opposite direction are inversely proportional. Figure 13–1 shows these relationships.

The branches in a parallel circuit respond similarly for current. As the resistance in a branch increases, the current decreases. Remember, though, the voltage across the branch remains the same as long as the source voltage does not change.

■ CURRENT DIVIDERS

So far, you have seen parallel circuits simply as separate parallel paths for current all returning to the same power source. Let's look at a two-branch parallel circuit in a different way. Figure 13–2 shows a circuit with two paths for current. You can see that the current divides at point A, travels through the two paths, and recombines at point B before returning to the source.

If the total current for this circuit were known but not the voltage, calculating the current through one of the resistors would be a bit involved. However, by simplifying the equations and relating one branch current to the total circuit current, you make the calculation easier.

The current through any branch of a parallel circuit is

$$I_1 = \frac{E_T}{R_1}$$

In addition, the total voltage in any circuit is

$$E_T = I_T \times R_T$$

Substituting total voltage into the current equation,

$$I_1 = I_T \frac{R_T}{R_1}$$

FIGURE 13–2   Current flow in a current divider circuit.

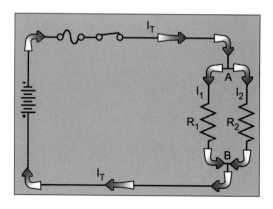

The general form of this equation is

$$I_x = I_T \frac{R_T}{R_x}$$

This equation can be used to calculate the current though any circuit branch by first calculating the total resistance (R_T) of the branch using the parallel resistance formulas you learned earlier and then using the values in Equation 1.

EXAMPLE 1

What are the currents through R_1 and R_2 in Figure 13–3?

Solution:
Using the current divider equation,

$$I_1 = I_T \frac{R_T}{R_1} = 1.5 \times \frac{17.53}{27} = 0.974 \text{ A}$$

$$I_2 = I_T \frac{R_T}{R_2} = 1.5 \times \frac{17.53}{50} = 0.526 \text{ A}$$

Now let's check the result using Ohm's law:

$$E_T = I_T \times R_T = 1.5 \times 17.53 = 26.295 \text{ V}$$

Since the voltage in a parallel circuit is the same across all the branches, solve for the current through R_1 and R_2 and see if the total current I_T equals the sum of the current through each of the resistances. The calculations are as follows:

$$I_1 = \frac{E_T}{R_1} = \frac{26.295}{27} = 0.974 \text{ A}$$

and

$$I_2 = \frac{E_T}{R_2} = \frac{26.295}{50} = 0.526 \text{ A}$$

FIGURE 13–3   Calculating currents in a current divider circuit.

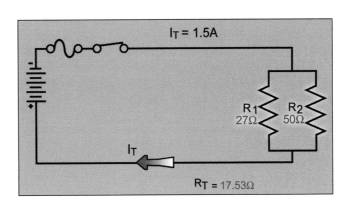

$I_T = 1.5A$

R1 27Ω   R2 50Ω

I_T

$R_T = 17.53\Omega$

The check values show that the original calculation was correct. The total current, I_T, can be found by adding the currents from each of the branches as follows:

$$I_T = I_1 + I_2 = .974 + .526 = 1.5 \text{ A}$$

■ LAW OF PROPORTIONALITY

As with voltage dividers, there is a Law of Proportionality that applies to current dividers. This law states that the amount of current passing through a branch of a current divider is proportional to the resistance of the other branch divided by the sum of the two branches times the total current. The following is a mathematical description of this law. From the previous equation,

$$I_1 = I_T \frac{R_T}{R_1}$$

Since:

$$R_T = \frac{R_1 R_2}{R_1 + R_2}$$

Substitute as follows:

$$I_1 = I_T \frac{R_T}{R_1} = I_T \frac{\dfrac{R_1 R_2}{R_1 + R_2}}{R_1} = I_T \frac{R_2}{R_1 + R_2}$$

Using the same approach, you can show that

$$I_2 = I_T \frac{R_1}{R_1 + R_2}$$

EXAMPLE 1

Calculate the currents through both branches of the circuit in Figure 13–3 using the Law of Proportionality.

$$I_1 = I_T \frac{R_2}{R_1 + R_2} = 1.5 \times \left(\frac{50}{27 + 50} \right) = 0.974 \text{ A}$$

and

$$I_2 = I_T \frac{R_1}{R_1 + R_2} = 1.5 \times \left(\frac{27}{27 + 50} \right) = 0.526 \text{ A}$$

Notice that both methods (current divider and Law of Proportionality) produced the same result as using Ohm's law.

■ SUMMARY

Current flows through parallel branches are inversely proportional to the amount of resistance in the branch: The higher the resistance, the lower the current. The amount of current can be calculated using the Law of Proportionality.

In circuits that have only two resistors or that are simplified to two resistors using equivalent circuits, the currents through each branch are proportional to the value of the resistance of the other branch divided by the sum of the branch resistances.

■ REVIEW QUESTIONS

1. In a parallel DC circuit, if you increase the resistance, what happens to the voltage drop?
2. In a parallel DC circuit, if you increase resistance, what happens to the value of current?
3. In a DC parallel circuit, current and resistance are _____ (directly/inversely) proportional.

4. In a DC parallel circuit, voltage and resistance are _____ (directly/inversely) proportional.
5. What is the constant in a parallel DC circuit?
6. What is the constant in a series DC circuit?

■ PRACTICE PROBLEM

1. Fill in the unknown quantities in the following problem. Use the Law of Proportionality.

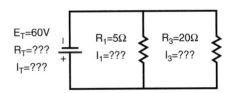

chapter 14

How to Calculate Power in a DC Parallel Circuit

■ **OUTLINE**

■ OVERVIEW

This chapter explains the theory and calculations of power in parallel circuits. Power calculations for parallel circuits are similar to the power calculations for series circuits. The total power in a parallel circuit is the sum of the power in the individual components of that circuit just as the total power in a series circuit is equal to the sum of the power in the individual components of that circuit.

Power calculations for these two types of circuits are also similar in that, in both circuits, the total power is the product of the source voltage and the total circuit current.

As you complete this chapter on parallel circuits, you should start to develop an understanding of the relationships between the variables in series and parallel circuits. This understanding will help you in the next series of lessons, which deal with combination circuits, and contain both series and parallel components. Equation 1 expresses these concepts mathematically:

$$P_L = P_1 + P_2 + P_3 + \ldots + P_n \tag{1}$$

One of the most important reasons for calculating values in electrical circuits is to determine the size and ratings of system components. For example, if a certain resistor is shown to dissipate 2 watts in the circuit, the electrician knows that the resistor will have to be rated 2 watts or higher.

■ OBJECTIVES

After completing this chapter, you should be able to:

1. Show the power required by individual components in a parallel circuit.
2. Calculate the total power consumed in a parallel circuit using the power consumed by individual components.
3. Calculate the total power consumed in parallel circuits from the source voltage and total current delivered to that circuit.
4. Determine power ratings of components in parallel circuits.

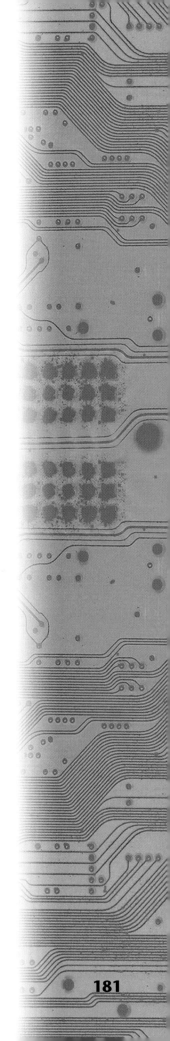

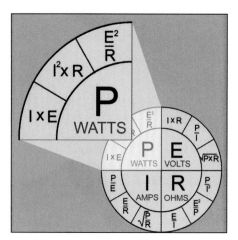

FIGURE 14–1   Ohm's law pie chart.

■ POWER IN PARALLEL

The equations for calculating power in parallel are the same as those used in series circuits. The pie chart for Ohm's law (Figure 14–1) shows the three equations you may use.

These equations are applied in the same manner as in a series circuit. The power for individual components can be calculated if any two of the three required values are known. The calculation for power of the total circuit is also the same. There are two basic methods for determining total power: directly or indirectly.

14.1 The Direct Method

Calculating power directly involves using one of three equations and replacing the symbols with total voltage (E_T), total current (I_T), or total resistance (R_T). The following examples will use Figure 14–2 to show the direct method for calculating power.

EXAMPLE 1

Calculate the total power in Figure 14–2.

Solution:
The values we have to calculate power are resistance and current. So, from Ohm's law,

$$P_T = I^2 R_T$$

Since the resistance in this equation is total resistance and this is a two-branch circuit, substitute the product over sum equation into the previous equation:

$$P_T = I^2 \frac{R_1 \times R_2}{R_1 + R_2} = 2^2 \left(\frac{150 \times 250}{150 + 250} \right) = 375 \text{ W}$$

EXAMPLE 2

Determine the source voltage in Figure 14–2.

Solution:
Rearranging the equation for power that includes voltage,

$$P_T = I_T \times E_T$$

FIGURE 14–2   Calculating power directly.

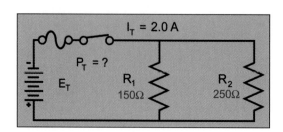

results in

$$E_T = \frac{P_T}{I} = \frac{375}{2} = 187.5 \text{ V}$$

14.2 The Indirect Method

The indirect method involves calculating the power for each individual component, then adding all the results to determine total power.

EXAMPLE 3

Use the indirect method and calculate total power for the circuit in Figure 14–3.

Solution:
The power for each component is determined using voltage and resistance:

$$P_1 = \frac{E_1^2}{R_1} = \frac{100^2}{150} = 66.67 \text{ W}$$

$$P_2 = \frac{E_2^2}{R_2} = \frac{100^2}{250} = 40 \text{ W}$$

So, the total power is

$$P_T = P_1 + P_2 = 66.67 + 40 = 106.67 \text{ W}$$

EXAMPLE 4

Is the fuse for this circuit (Figure 14–3) rated properly? Remember that the fuse is the circuit safety device. If the circuit uses too much power, the current will be too high and the fuse will open, stopping the circuit current.

FIGURE 14–3   Calculating power indirectly.

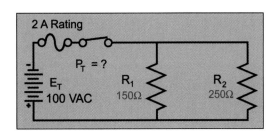

Solution:

Since the total power and voltage is known, the equation to re-arrange is

$$P = I \times E \Rightarrow I = \frac{P}{E}$$

$$I = \frac{106.67}{100} = 1.067 \text{ A}$$

These calculations show that the fuse will allow the normal load current to flow without blowing. Of course, a 1.5-ampere fuse would also work. The exact one to choose is based on other issues.

■ SUMMARY

There are several ways to calculate the total power consumed in a circuit or in the components of a circuit. The total power of a circuit can be calculated directly by using the total source voltage (E_T), total resistance (R_T), and/or total current (I_T).

The total power consumed by a circuit can also be calculated indirectly by first computing the power consumed in each component and then adding all these loads together. These relationships need to be thoroughly understood, as they are part of the electrical theory foundation that every electrical worker relies on day in and day out.

After the various power requirements are calculated, circuit elements can then be sized to be certain that they will properly carry and/or dissipate the currents and powers.

■ REVIEW QUESTIONS

1. The total power in a series circuit is equal to the _____ of the power in the individual components of that circuit.

2. The total power in a DC parallel circuit is equal to the _____ of the power consumed in the individual components of that circuit.

3. If the power is known across each branch of a parallel circuit, what is the formula for solving total power? Is this formula the same for a series DC circuit?

4. If total voltage and current are known, what is the formula for solving total power?

5. If total voltage and total resistance are known, what is the formula for solving total power?

6. If total current and total resistance are known, what is the formula for solving total power?

■ PRACTICE PROBLEMS

1. A DC parallel circuit has a source voltage of 30 volts, and there are two parallel branches. Branch 1 has a resistance of 6 ohms, and branch 2 has a resistance of 3 ohms.
   a. Draw the circuit.
   b. Find the current drawn across each load.
   c. Find the total current.
   d. Find the total resistance.
   e. Find the power across each load.
   f. Find the total power.

2. Using the circuit values from Exercise 1:
   a. Solve for the power across each branch of the circuit.
   b. Solve for total power using the following formulas:

$$P = \frac{E^2}{R} \text{ and } P = I^2 R$$

PART

4

DC COMBINATION CIRCUITS

chapter 15

Understanding Resistance in Combination Circuits

■ OUTLINE

■ OVERVIEW

All the chapters you have studied to this point were concerned with either a series circuit or a parallel circuit. This chapter introduces combination circuits. A combination circuit is one in which both series and parallel conditions exist. Few of the practical circuits that you will encounter in your career will be purely series circuits or purely parallel circuits. Most electrical systems are made up of combination circuits—the combination of series and parallel components.

In this chapter, you will learn how to reduce or combine resistances in combination circuits so that the resulting equivalent resistance appears to be nothing more than a single series resistor. An example of this is combining three parallel branch resistances into one equivalent resistance and adding this equivalent resistance to the rest of the circuit to calculate the total circuit resistance. This is illustrated in Figure 15–1, which shows one resistor (R_4) in series with the parallel combination of three other resistors (R_1, R_2, R_3).

■ OBJECTIVES

After completing this chapter, you should be able to:

1. Identify circuits that are classified as combination or series parallel circuits.
2. Analyze components in a combination circuit to determine whether they are connected in series or parallel with other components.
3. Apply the rules you learned for series and parallel resistors to reduce a circuit to its equivalent resistance.

■ GLOSSARY

Combination circuit   A circuit made up of both parallel and series elements.

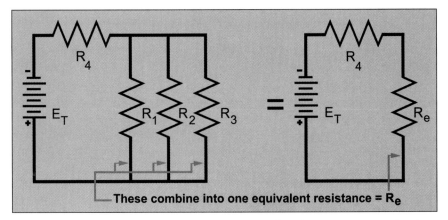

FIGURE 15–1   Example of series and parallel elements in one circuit (combination circuit).

■ RESISTANCE IN COMBINATION CIRCUITS

15.1 Reducing a Circuit

Figure 15–2 is a simple combination circuit. The first step to analyze this circuit is to reduce (simplify) the circuit as much as possible. Each section to be reduced will be a group of two or more resistors with the equivalent resistance results taking the place of the group. The parallel parts of the circuit are analyzed first.

Resistors R_2 and R_3 are in parallel. First, find the equivalent resistance for R_2 and R_3 by using the product over sum equation:

$$R_e = \frac{R_2 \times R_3}{R_2 + R_3}$$

where R_e is the equivalent resistance:

$$R_e = \frac{100 \times 200}{100 + 200} = 66.67 \ \Omega$$

With this result, redraw the circuit and replace the two parallel resistors with their "equivalent" single resistor as shown in Figure 15–3.

Now the circuit in Figure 15–2 has been reduced to a simple series circuit with two resistors in series. The total resistance is

$$R_T = R_1 + R_e = 250 + 66.67 = 316.67 \ \Omega$$

15.2 Solving More Complex Combination Circuits

Reducing a more complex circuit to its equivalent series resistor is performed much the same as previously. Table 15–1 shows the general steps involved.

The circuit shown in Figure 15–4 will be reduced to its equivalent series resistor. Each section will be reduced individually and the cir-

FIGURE 15–2   Simple combination circuit.

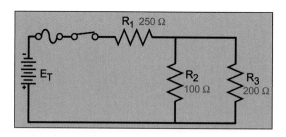

FIGURE 15–3   Circuit of Figure 15–2 showing equivalent resistance (R_e) of R_2 and R_3.

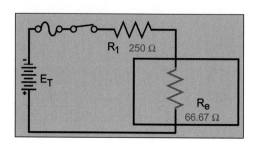

Table 15–1   **Suggested Steps to Reduce Complex Combination Circuits to Equivalent Resistance Value**

Step	Description
1	Reduce one part of the circuit at a time.
2	Redraw the circuit, exchanging the equivalent circuit for the original resistors. This helps you visualize the next required step.
3	Ensure that all series resistors have been combined before a parallel portion is reduced.
4	Combine parallel portions to a single resistor.
5	Repeat combining equivalent resistors until all portions are reduced to one equivalent resistance.

FIGURE 15–4   Complex combination circuit.

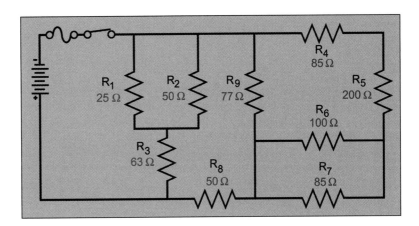

cuit redrawn to show the equivalent circuit until only one resistor remains. The first task is to determine which series resistors may be combined. In this circuit, only R_4 and R_5 may be combined; all the rest have parallel components. From Figure 15–5,

$$R_{4,5} = R_4 + R_5 = 85 + 200 = 285 \ \Omega$$

The circuit may then be redrawn (Figure 15–6) with the equivalent resistance drawn in place of R_4 and R_5.

The next step is to combine any parallel portions of the circuit. The parallel parts of the circuit are $R_1 - R_2$ and $R_6 - R_7$ (Figures 15–7 and 15–8). Using the product over sum equation,

$$R_{1,2} = \frac{R_1 \times R_2}{R_1 + R_2} = \frac{25 \times 50}{25 + 50} = 16.67 \ \Omega$$

$$R_{6,7} = \frac{R_6 \times R_7}{R_6 + R_7} = \frac{100 \times 85}{100 + 85} = 45.95 \ \Omega$$

Now redraw the circuit, combining the two parallel branch circuits into their respective equivalent resistances (Figure 15–9).

Next, since $R_{1,2}$ and R_3 are in series and $R_{4,5}$ and $R_{6,7}$ are in series, they can be added directly. See Figures 15–10 and 15–11. The result is

$$R_{1,2,3} = R_{1,2} + R_3 = 16.67 \ \Omega + 63 \ \Omega = 79.67 \ \Omega$$

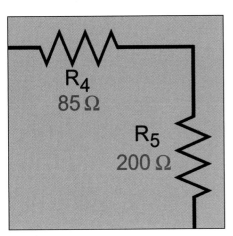

FIGURE 15–5   Resistors R_4 and R_5 from Figure 15–4.

FIGURE 15–6   Figure 15–4 redrawn with the equivalent resistance for R_4 and R_5.

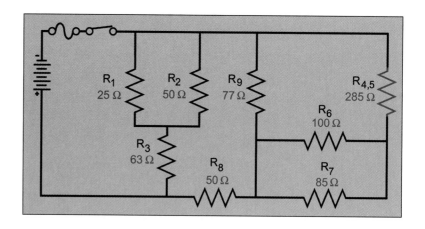

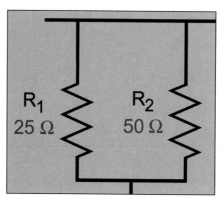

FIGURE 15–7   Reducing R_1 and R_2 to a single equivalent resistance.

FIGURE 15–8   Reducing R_6 and R_7 to a single equivalent resistance.

FIGURE 15–9   Figure 15–6 redrawn with equivalent resistances for R_1, R_2, R_6, and R_7.

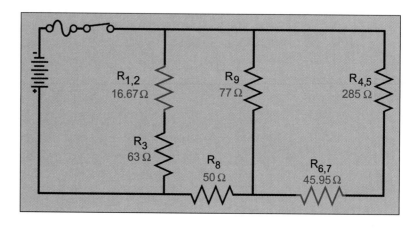

and

$$R_{4,5,6,7} = R_{4,5} + R_{6,7} = 285 \ \Omega + 45.95 \ \Omega = 330.95 \ \Omega$$

The result is redrawn and shown in Figure 15–12.

$R_{4,5,6,7}$ and R_9 are in parallel (Figure 15–13). The product over sum equation can be used as follows:

$$R_{4,5,6,7,9} = \frac{R_9 \times R_{4,5,6,7}}{R_9 + R_{4,5,6,7}} = \frac{77 \times 330.95}{77 + 330.95} = 62.47 \ \Omega$$

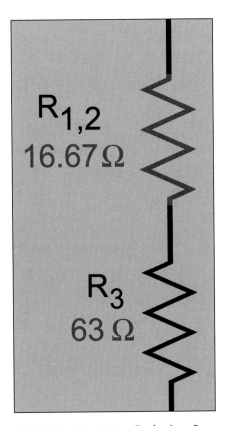

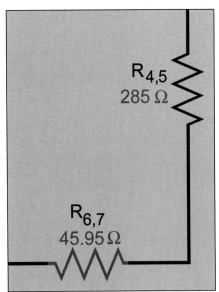

FIGURE 15–11 Reducing $R_{4,5}$ and $R_{6,7}$ to an equivalent resistance.

FIGURE 15–10 Reducing $R_{1,2}$ and R_3 to an equivalent resistance.

FIGURE 15–12 Figure 15–6 redrawn with equivalent resistance for $R_{1,2,3}$ and $R_{4,5,6,7}$.

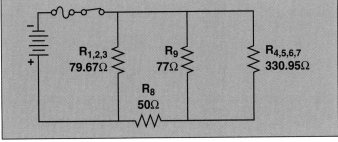

The redrawn circuit is shown in Figure 15–14.

Figure 15–15 shows the series combination of resistors $R_{4,5,6,7,9}$ and R_8:

$$R_{4,5,6,7,8,9} = R_8 + R_{4,5,6,7,9} = 50\ \Omega + 62.47\ \Omega = 112.47\ \Omega$$

The result (Figure 15–16) is a simple parallel circuit with two resistors. The product over sum equation is used one final time to create the final result shown in Figure 15–17:

$$R_{1,2,3,4,5,6,8,9} = \frac{R_{1,2,3} \times R_{4,5,6,7,8,9}}{R_{1,2,3} + R_{4,5,6,7,8,9}} = \frac{79.67 \times 112.47}{79.67 + 112.47} = 46.64\ \Omega$$

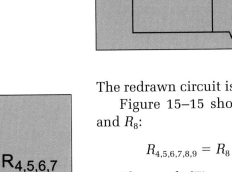

FIGURE 15–13 Reducing R_9 and $R_{4,5,6,7}$ to an equivalent resistance.

This is the equivalent resistance from Figure 15–4. The nine resistors from the original circuit appear as one single resistive source as shown in Figure 15–17.

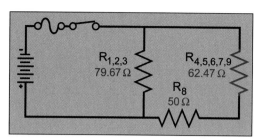

FIGURE 15–14   Original circuit of Figure 15–4 redrawn for $R_{1,2,3}$, R_8, and $R_{4,5,6,7,9}$.

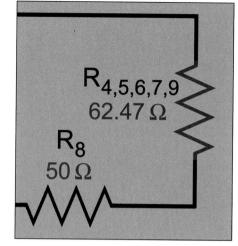

FIGURE 15–15   Reducing $R_{4,5,6,7,9}$ and R_8 to an equivalent resistance.

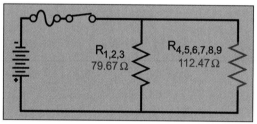

FIGURE 15–16   Original circuit of Figure 15–4 reduced to parallel equivalent resistances.

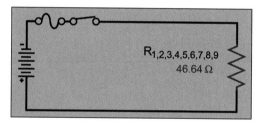

FIGURE 15–17   Final equivalent resistance for Figure 15–4.

■ SUMMARY

Combination circuits are those in which both series and parallel components exist. Using the techniques learned in the chapters on series and parallel circuits, the resistive components in combination circuits can be combined and redrawn until only one equivalent resistor remains. The equivalent resistor is the resistive component fed by the voltage source and determines the current characteristics for the entire circuit.

■ REVIEW QUESTIONS

1. Define a combination circuit.
2. What is the objective for solving total resistance in a combination circuit?

3. The equivalent resistance of a combination circuit, along with the source voltage, determines the total current flow (I_T) in the circuit. Discuss this in terms of the value of reducing combination circuits to equivalent resistances.

PRACTICE PROBLEM

1. Solve for R_T in the following three circuits.

a.

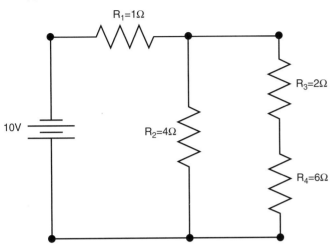

c.

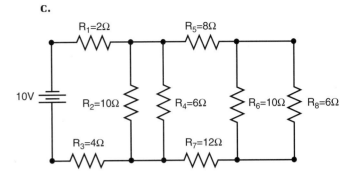

b.

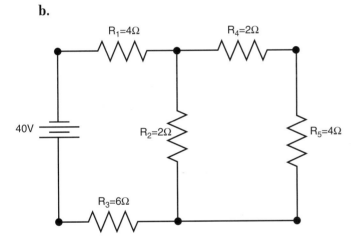

chapter 16

How Current Reacts in DC Combination Circuits

■ OUTLINE

OVERVIEW

In chapter 15, you were introduced to combination circuits and calculating the equivalent resistance of the circuit. This chapter continues with combination circuits and explores current characteristics in these types of circuits. Using the information learned in chapter 15, we will derive total current, individual series, and branch currents and show how to calculate the current through any single component in the circuit.

OBJECTIVES

After completing this chapter, you should be able to:

1. Identify alternative current paths in combination or series parallel circuits.
2. Determine which components will carry total circuit current in combination circuits.
3. Apply Ohm's law to determine the current through any branch or component of a combination circuit.

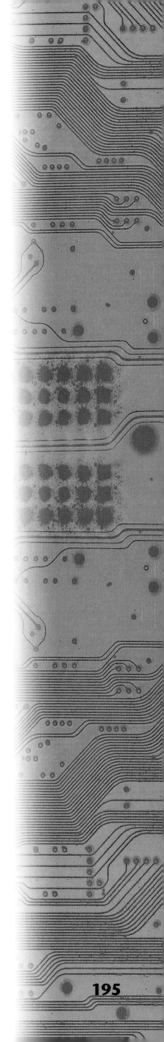

■ CURRENT IN COMBINATION CIRCUITS

16.1 Background

Combination circuits are circuits in which both series and parallel conditions exist. A series circuit, as you know, is one in which there is only one path for current. Parallel circuits have multiple paths. To determine the type of circuit, trace the path for current through the components using the schematic drawing. If it appears that the current can take more than one path, then the circuit is not series.

When you studied current in earlier chapters, you learned that in purely series circuits, all components have the same amount of current. In parallel circuits, the total current is equal to the sum of all the branch currents. The total current in a combination circuit can be calculated once the total equivalent resistance is known.

16.2 Current through a Simple Combination Circuit

Look at Figure 16–1. This is the same combination circuit from Figure 15–1. To determine the total current through the circuit, first determine the equivalent resistance. In chapter 15, it was shown that the equivalent resistance is equal to the resistance of R_1 plus the equivalent resistance of $R_{2,3}$. The resistance of $R_{2,3}$ is calculated using the product over sum equation for parallel resistors. So,

$$R_T = R_e = R_1 + R_{2,3} = R_1 + \frac{R_2 \times R_3}{R_2 + R_3}$$

$$R_e = 250 + \frac{100 \times 200}{100 + 200} = 250 + 66.67 = 316.67 \ \Omega$$

The circuit can now be redrawn as in Figure 15–2 (see Figure 16–2).
The total current can now be calculated using Ohm's law:

$$I_T = \frac{E_T}{R_e} = \frac{100}{316.67} = 0.316 \ A$$

With the value of total current calculated, look back at Figure 15–1. A current of 0.316 A is flowing through the parallel portion of the circuit. The parallel portion is a current divider. Each branch of the parallel portion of the circuit has a current flow value that, when added together, equals the total current flow through the branch.

FIGURE 16–1   Simple combination circuit.

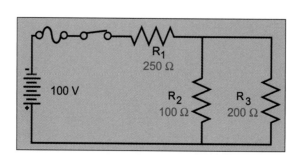

FIGURE 16–2 Equivalent resistance of combination circuit.

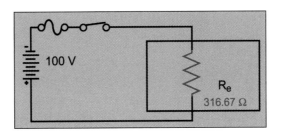

The total amount of current flowing through the circuit is 0.316 A because every combination circuit can be reduced to an equivalent circuit and the current values in a series circuit are the same throughout the circuit. Since the total current equals the current at any point in a series circuit, look again at Figure 15–1. Resistor R_1 and the combined group or the equivalent resistors R_2 and R_3 are in series with each other. Now look at the current values relative to the equivalent circuits:

$$I_T = I_1 = I_{2,3} = 0.316 \text{ A}$$

This is an important concept to remember. Since the current value through the parallel branch $R_{2,3}$ is known, the voltage drop across the $R_{2,3}$ branch can be calculated using Ohm's law as follows:

$$E_{2,3} = I \times R_{2,3} = 0.316 \times 66.67 = 21.07 \text{ V}$$

The voltage supplied to the parallel branch $R_{2,3}$ is known and the voltage is the same across all branches in a parallel circuit; therefore, the current through each resistance of the parallel branch can be calculated using Ohm's law as follows:

$$I_{R_2} = \frac{E_{2,3}}{R_2} = \frac{21.07}{100} = 0.2107 \text{ A}$$

$$I_{R_3} = \frac{E_{2,3}}{R_3} = \frac{21.114}{200} = 0.1053 \text{ A}$$

Check the answer. $I_{2,3} = 0.316$ A. Add the value calculated for each individual branch and see if it equals 0.316 A:

$$I? = I_2 + I_3 = .2107 + 1.053 = 0.316 \text{ A}$$

This proves that the current through each of the branches in a parallel circuit, when added together, equals the total current through the branch.

As discussed in chapter 13, you can also use the Law of Proportionality to find the current through each of the branches. The calculations are shown here:

$$I_2 = I_T \times \left(\frac{R_3}{R_2 + R_3} \right) = 0.316 \times \left(\frac{200}{100 + 200} \right) = 0.2107$$

$$I_3 = I_T \times \left(\frac{R_2}{R_2 + R_3} \right) = 0.316 \times \left(\frac{100}{100 + 200} \right) = 0.1053$$

16.3 Solving Complex Circuits

While the Law of Proportionality works on circuits where the parallel branches are well defined, circuits that are more complex and have many branches in combination must be solved using Ohm's law. Figure 16–3 (the same as Figure 15–3) provides a more complex circuit to calculate the currents through each individual resistor.

EXAMPLE 1

What is the total current in Figure 16–3?

Solution:

From working with this problem in a previous lesson, the total voltage $E_T = 200$ V and the total resistance $R_T = 46.64$ Ω. By knowing these two variables, solve for the total current, I_T, by using Ohm's law. The total equivalent resistance circuit for Figure 16–3 can be seen in Figure 16–4.

$$I_T = \frac{E_T}{R_e} = \frac{200}{46.64} = 4.29 \text{ A}$$

EXAMPLE 2

What are the current values through each resistor 1 to 9?

Solution:

The value of currents through each respective resistor must also be calculated using Ohm's law. Work backward from the total equivalent resistance, $R_T = 46.64$ Ω, as shown in Figure 16–4 and calculate the current through each branch. Figure 16–5 shows the first expanded step of the circuit. Figure 16–5 is the same as Figure 15–16.

Figure 16–5 is an expansion of Figure 16–4 and shows how the single resistance breaks into a simple parallel circuit with two

FIGURE 16–3 Complex combination circuit.

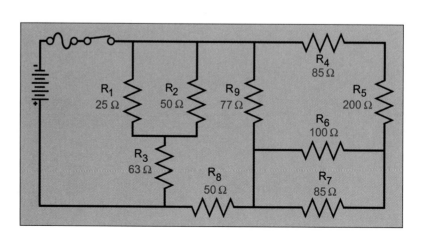

FIGURE 16–4   Total equivalent circuit.

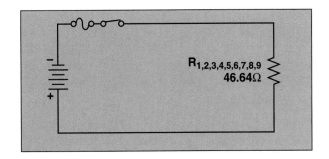

FIGURE 16–5 Complex combination circuit showing two branches.

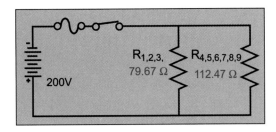

branches. The current can be calculated directly using Ohm's law since the voltage source is impressed across both branches:

$$I_{1,2,3} = \frac{E_T}{R_{1,2,3}} = \frac{200}{79.67} = 2.51 \text{ A}$$

$$I_{4,5,6,7,8,9} = \frac{E_T}{R_{4,5,6,7,8,9}} = \frac{200}{112.47} = 1.78 \text{ A}$$

Checking the results, you see that the sum of the two branch currents is equal to the total current:

$$I_T = I_{1,2,3} + I_{4,5,6,7,8,9} = 2.51 + 1.78 = 4.29 \text{ A}$$

Separating $R_{1,2,3}$ back into its individual resistors results in the branch shown in Figure 16–6. From the prior calculation, 2.51 A are flowing though this branch. The 2.51 A are shared by resistors R_1 and R_2 because they are in parallel. The total 2.51 A flow through R_3

FIGURE 16–6   First branch from Figure 16–3.

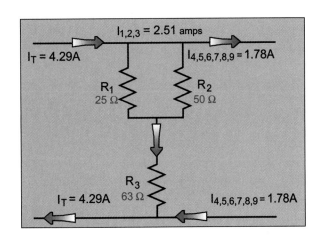

since it is in series with the parallel branch $R_{1,2}$. Since R_1 and R_2 are a current divider, the Law of Proportionality equation can be used:

$$I_1 = I_{1,2,3} \times \left(\frac{R_2}{R_1 + R_2}\right) = 2.51 \times \left(\frac{50}{25 + 50}\right) = 1.67 \text{ A}$$

$$I_2 = I_{1,2,3} \times \left(\frac{R_1}{R_1 + R_2}\right) = 2.51 \times \left(\frac{25}{25 + 50}\right) = 0.84 \text{ A}$$

The current through R_3 is simply the branch current because it is a series resistor in the branch:

$$I_3 = 2.51 \text{ A}$$

From Figure 15–12, you see that $R_{4,5,6,7,8,9}$ can be drawn as two parallel resistors in series with one resistor (see Figure 16–7).

The current through R_8 can be stated directly since it is a series resistor in the branch:

$$I_8 = 1.78 \text{ A}$$

The current through R_9 is calculated using the Law of Proportionality:

$$I_9 = I_{4,5,6,7,8,9} \times \left(\frac{R_{4,5,6,7}}{R_9 + R_{4,5,6,7}}\right) = 1.78 \times \left(\frac{330.95}{77 + 330.95}\right) = 1.44 \text{ A}$$

This means that the remainder of the 1.78 A flows through $R_{4,5,6,7}$:

$$I_{4,5,6,7} = I_{4,5,6,7,8,9} - I_9$$
$$I_{4,5,6,7} = 1.78 - 1.44 = 0.34 \text{ A}$$

From Figure 16–8, you can see that R_4 and R_5 are in series with the branch, so they have the same current value as the whole branch:

$$I_4 = I_5 = 0.34 \text{ A}$$

R_6 and R_7 are parallel resistors, so the Law of Proportionality applies:

$$I_6 = I_{4,5,6,7} \times \left(\frac{R_7}{R_6 + R_7}\right) = 0.34 \times \left(\frac{85}{100 + 85}\right) = 0.16 \text{ A}$$

$$I_6 = I_{4,5,6,7} \times \left(\frac{R_6}{R_6 + R_7}\right) = 0.34 \times \left(\frac{100}{100 + 85}\right) = 0.18 \text{ A}$$

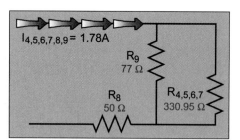

FIGURE 16–7   Second branch from Figure 16–3.

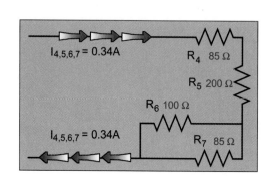

FIGURE 16–8   Current through third branch of Figure 16–3.

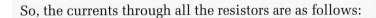

So, the currents through all the resistors are as follows:

$I_1 = 1.67$

$I_2 = 0.84$

$I_3 = 2.51$

$I_4 = 0.34$

$I_5 = 0.34$

$I_6 = 0.16$

$I_7 = 0.18$

$I_8 = 1.78$

$I_9 = 1.44$

Note that adding the current of all the resistors does not give you the total current (I_T) for the circuit. This is because some of the resistors are in parallel.

■ SUMMARY

The current through a combination circuit is calculated first by reducing the circuit to its simplest form, then applying either Ohm's law or the Law of Proportionality. Once the total current is known, work backward, expanding the equivalent resistors into their original form and calculating current as each branch is expanded.

■ REVIEW QUESTIONS

1. A circuit that has both series and parallel branches is called what type of circuit?

2. Describe generally how the current will divide among the various branches of a parallel circuit. Use both an Ohm's law approach and a Law of Proportionality approach.

3. In a parallel circuit, _____ is the same throughout the circuit.

4. In a series circuit, _____ is the same throughout the circuit.

■ PRACTICE PROBLEM

1. Solve for the current flow in each resistor of Exercises 1, 2, and 3 in chapter 15.

chapter

17

How Voltage Functions in DC Combination Circuits

■ OUTLINE

■ OVERVIEW

In this, the third chapter on combination circuits, you will study voltage relationships. In chapter 16, you learned that in order to solve for total or branch currents in combination circuits, you had to solve for current, resistance, and/or voltage throughout the circuit. In this chapter, the emphasis is placed on voltage calculations in similar combination circuits.

The main tools that an electrician uses in solving problems such as these are formulas regarding series and parallel circuits, combined with Ohm's law. Using these principles, you will be able to solve the problems presented in this chapter with little difficulty.

■ OBJECTIVES

After completing this chapter, you should be able to:

1. Identify the components in combination circuits that have the same voltage dropped across them.
2. Determine how voltage drops are distributed in a combination circuit to equal the source value.
3. Apply Ohm's law to determine the voltage drop across any component in a combination circuit as well as the circuit's applied voltage.

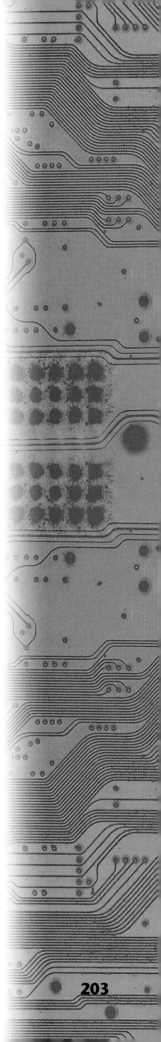

■ CALCULATING VOLTAGE DROPS IN COMBINATION CIRCUITS

17.1 Starting Principles

The two previous chapters provide the beginning steps in determining circuit voltages:

1. Calculate the equivalent resistance
2. Calculate circuit currents
3. Calculate voltage drops using Ohm's law

As you study and practice the problems in this chapter, remember that actual application goes well beyond the diagrams shown. Always make every effort to relate the lesson material to what you have experienced or observed on the job. For example, consider three lamps connected in parallel to a 120-volt supply. Your first thought relative to this circuit might resemble the physical circuit shown in Figure 17–1. A simple schematic equivalent circuit for the three lamps is shown in Figure 17–2.

In fact, if you consider the voltage drop of the conductors, Figure 17–2 is not entirely accurate. Earlier in this course, you learned that the conductors (wires) would also have resistance and a subsequent voltage drop. We have ignored those drops before; however, the diagram should really look more like Figure 17–3. This new, more accurate diagram is in fact a combination circuit, with series and parallel components.

FIGURE 17–1   Lamps in parallel.

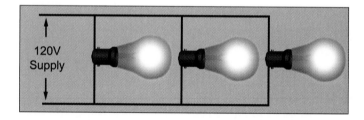

FIGURE 17–2   Equivalent resistive circuit for lamps in parallel.

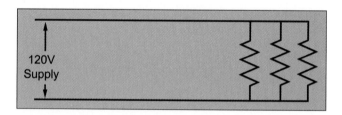

FIGURE 17–3   Equivalent resistive circuit for lamps in parallel including wire resistance.

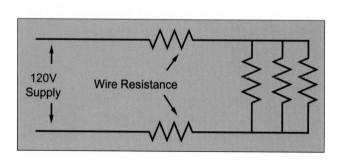

17.2 Voltage across a Simple Combination Circuit

Figure 17–4 is the same simple combination circuit from chapter 15 that you have used before. To determine the voltage drop across each resistor, the equivalent resistance must be calculated first. It has been shown that the equivalent resistance is equal to the resistance of R_1 plus the equivalent resistance of $R_{2,3}$:

$$R_T = R_1 + \left(\frac{R_2 \times R_3}{R_2 + R_3}\right) = 250 + \left(\frac{100 \times 200}{100 + 200}\right) = 316.67 \ \Omega$$

The circuit is redrawn as in Figure 17–5 with R_e shown as equivalent resistance.

The total current can now be calculated using Ohm's law:

$$I_T = \frac{E_T}{R_e} = \frac{100}{316.67} = 0.316 \text{ A}$$

With the value of total current calculated, look back at Figure 17–4. You now know that 0.316 A are flowing through R_1 to the parallel portion of the circuit. The voltage drop across R_1 can be calculated using Ohm's law:

$$E_1 = I_T R_1 = (0.316)(250) = 79 \text{ V}$$

You learned earlier that the sum of the voltages in a series circuit is equal to the total or source voltage. Resistor R_1 drops 79 volts of the source voltage of 100 volts. The remaining voltage must be dropped across the equivalent resistance of R_2 and R_3 in parallel (see Figure 17–4). The voltage is calculated as follows:

$$E_{equiv.} = E_T - E_1 = 100 - 79 = 21 \text{ V}$$

FIGURE 17–4 Simple combination circuit.

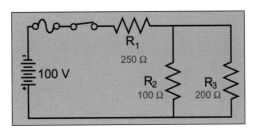

FIGURE 17–5 Equivalent resistance of combination circuit shown in Figure 17–4.

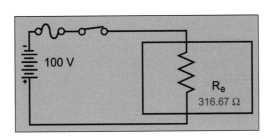

Since the equivalent circuit is made up of two resistors in parallel, R_2 and R_3, the voltage drop across one branch of the parallel circuit is equal to the voltage drop across the other branch:

$$E_{equiv.} = E_2 = E_3 = 21 \text{ V}$$

17.3 Solving Complex Circuits

In chapter 16, you used the circuit shown in Figure 17–6 and derived the current through each resistor. You reduced the resistance to an equivalent resistance (Figure 15–17) and calculated the total current as follows:

$$I_T = \frac{E_T}{R_e} = \frac{200}{46.64} = 4.29 \text{ A}$$

Working backward, you expanded the equivalent resistances into their individual circuit component values and stopped when you had series equivalent resistors for which the current could be calculated. In the first step, you expanded the circuit into two parallel resistors (see Figure 17–7) and found the current through each to be as follows:

$$I_{1,2,3} = \frac{E_T}{R_{1,2,3}} = \frac{200}{79.67} = 2.51 \text{ A}$$

and

$$I_{4,5,6,7,8,9} = \frac{E_T}{R_{4,5,6,7,8,9}} = \frac{200}{112.47} = 1.78 \text{ A}$$

FIGURE 17–6   Complex combination circuit.

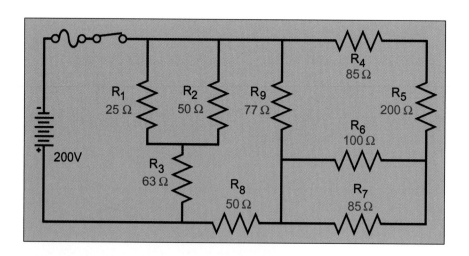

FIGURE 17–7   Complex combination circuit showing two branches.

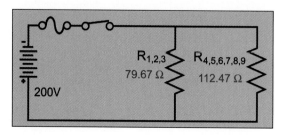

FIGURE 17-8   First branch from Figure 17-6.

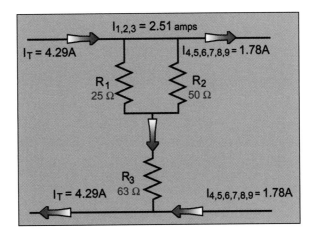

You then expanded the first branch into its individual resistors (see Figure 17-8). You used Ohm's law and the Law of Proportionality to calculate the current through each individual resistor. The results are

$$I_{R_1} = I_{1,2,3} \times \left(\frac{R_2}{R_1 + R_2}\right) = 2.51 \times \left(\frac{50}{25 + 50}\right) = 1.67 \text{ A}$$

$$I_{R_2} = I_{1,2,3} \times \left(\frac{R_1}{R_1 + R_2}\right) = 2.51 \times \left(\frac{25}{25 + 50}\right) = 0.84 \text{ A}$$

The current through R_3 is simply the branch current because it is a series resistor in the branch:

$$I_{R_3} = 2.51 \text{ A}$$

You then expanded $R_{4,5,6,7,8,9}$ into two parallel resistors in series with one resistor (see Figure 17-9). The current through series resistor R_8 is

$$I_8 = 1.78 \text{ A}$$

and the current through R_9, using the Law of Proportionality, is

$$I_{R_9} = I_{4,5,6,7,8,9} \times \left(\frac{R_{4,5,6,7}}{R_{4,5,6,7} + R_9}\right) = 1.78 \times \left(\frac{330.95}{330.95 + 77}\right) = 1.44 \text{ A}$$

The remainder of the current flowed through $R_{4,5,6,7}$ (see Figure 17-10):

$$I_{4,5,6,7} = I_{4,5,6,7,8,9} - I_{R_9} = 1.78 - 1.44 = 0.34 \text{ A}$$

FIGURE 17-9   Second branch from Figure 17-6.

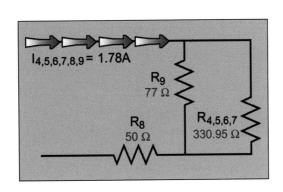

FIGURE 17–10 Current through third branch of Figure 17–6.

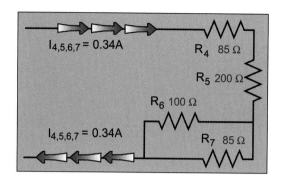

Since R_4 and R_5 are in series, they have the same current value as the whole branch:

$$I_4 = I_5 = 0.34 \text{ A}$$

The current through parallel resistors R_6 and R_7 is derived using Ohm's law. The equivalent resistance of R_6 and R_7 is calculated as

$$R_{6,7} = \frac{R_6 \times R_7}{R_6 + R_7} = \frac{100 \times 85}{100 + 85} = 45.95 \ \Omega$$

Then the voltage drop across the equivalent $R_{6,7}$ branch must be calculated. You know from prior calculations that $I_{4,5,6,7} = 0.34$ A; therefore, the current flow through $R_{6,7}$ must also be 0.34 A (see Figure 17–11):

$$E_{6,7} = I_{6,7} \times R_{6,7} = 0.34 \times 45.95 = 15.62 \text{ V}$$

Therefore, the voltage drop across R_6 equals the voltage drop across R_7, which equals 15.62 V because R_6 and R_7 are in parallel. With the voltage known, you can calculate the current through R_6 and R_7 (see Figure 17–12):

$$I_{R_6} = \frac{E_{6,7}}{R_6} = \frac{15.62}{100} = 0.1562 \text{ A}$$

$$I_{R_7} = \frac{E_{6,7}}{R_7} = \frac{15.62}{85} = 0.1838 \text{ A}$$

To check your answers, add the currents through R_6 and R_7:

$$I_{6,7} = I_6 + I_7 = .1562 + .1838 = .34$$

FIGURE 17–11 Voltage drop across $R_{6,7}$.

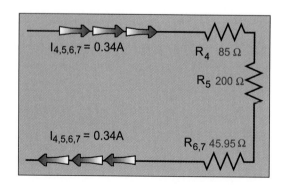

FIGURE 17–12 Current through $R_{6,7}$.

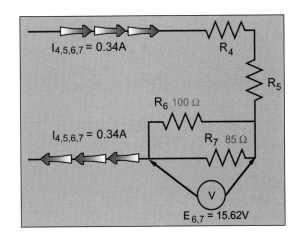

Table 17–1 Voltage Drops for Combination Circuit Resistors

Resistor	Resistance	Current	Voltage
R_1	25	1.67	41.84
R_2	50	0.84	41.84
R_3	63	2.51	158.13
R_4	85	0.34	28.90
R_5	200	0.34	68.00
R_6	100	0.1562	15.62
R_7	85	0.1837	15.62
R_8	50	1.78	89.00
R_9	77	1.44	110.88

Now placing all known values into the chart and calculating the remaining resistors by applying Ohm's law produces Table 17–1.

SUMMARY

To determine the voltage drop across individual resistors in a combination circuit, the equivalent resistance must be calculated first, then the total circuit current. The current is then applied to series and parallel portions of the circuit. Use Ohm's law and the Law of Proportionality to determine the current through each individual resistor. With the current for each component calculated, use Ohm's law again to get voltage.

REVIEW QUESTIONS

1. The first thing you need to calculate in combination circuits is the _____.
2. What are the main tools used in solving combination circuits?
3. What is calculated after equivalent resistance or total resistance?
4. What are the final steps in solving combination circuits? In solving branch voltage drops?

■ PRACTICE PROBLEM

1. Using the circuit in the following figure, calculate the currents and voltages in each circuit element.

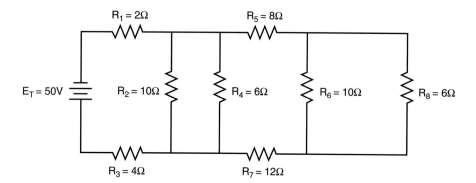

chapter 18

How to Calculate Power in DC Combination Circuits

■ OUTLINE

■ OVERVIEW

A s with all other types of circuits, the power utilized in a combination circuit is the sum of the power consumed by each of the individual components. Each component adds a portion to the total power. You have worked in prior lessons to systematically reduce the various components in a combination circuit down to an "equivalent resistance" that is representative of the various components regardless of whether they are connected in series or parallel in the circuit. This equivalent resistance can also be used to calculate the total power of the combination circuit. When the total equivalent resistance is used to find the power, the value will be equal to the sum of the power consumed by the individual components.

In this chapter you will continue to develop additional skills in analyzing combination circuits and specifically in determining the power used in these circuits. You will now learn how to determine the power requirements of a combination circuit. In addition, you will continue to practice the same systematic approach to solve for the various unknown values by analyzing the circuit and working with the information that is given you. These values, when inserted into the various formulas, will provide insight into all the possible circuit parameters. In addition, you will also be using known power values to calculate for voltage and current, working backward to solve for the unknown. Again, the key to solving all combination circuits is building confidence in analyzing the circuit for what is known and then systematically solving for the unknowns by reducing the various resistances to an equivalent resistance value.

■ OBJECTIVES

After completing this chapter, you should be able to:

1. Calculate the power dissipated in components of combination circuits.
2. Determine the power requirements of a circuit when it is necessary to do so.
3. Solve for other variables in a combination circuit when power is known.

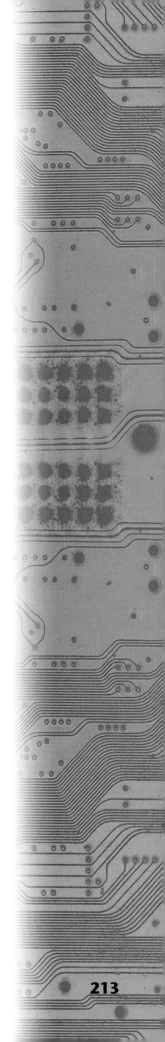

■ POWER IN COMBINATION CIRCUITS

18.1 Calculating Power

As you learned in previous chapters, the total power consumed by a series or parallel circuit is equal to the sum of the power consumed by each of the circuit's components. The same is true of the combination circuit. Often only one parameter, such as the resistance, the voltage, or the current, will be given. Your job is to reduce and combine the various resistances to find the second parameter. Often this requires the systematic reduction of the circuit to a single equivalent resistance value. Then you work back in the opposite direction to provide the missing values for the various components using Ohm's law. Remember, you cannot find the power until you have two of the three parameters for each of the components.

The total power is derived by adding all the individual powers from each of the components:

$$P_T = P_1 + P_2 + P_3 + \ldots P_n$$

Calculate the power dissipated by each component and the total power of the combination circuit in Figure 18–1. The first step is to review the circuit, analyzing for what is known and what is not known and comparing what you see to what you are solving for. In Figure 18–1 there are no components that have two values from which you can work. This means that you will have to begin by combining circuit components to solve for the necessary second variable.

You could use the equation $P_T = E_T \times I_T$ to calculate the total power; however, Figure 18–1 does not have the value for I_T. The first step then is to find I_T. To calculate the value for I_T, you need the value for the total resistance, R_T. Knowing R_T allows you calculate I_T by using the formula

$$I_T = \frac{E_T}{R_T}$$

You previously calculated R_T in this circuit back in chapter 15. You start by combining resistors R_2 and R_3:

$$R_{2,3} = \frac{R_2 \times R_3}{R_2 + R_3} = \frac{100 \times 200}{100 + 200} = 66.67 \ \Omega$$

FIGURE 18–1   Simple combination circuit.

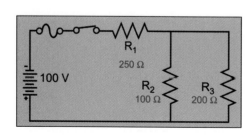

The second step is to reduce series resistor R_1 and equivalent resistor $R_{2,3}$ to a single value. This gives R_T. Since this is a series circuit, R_T is calculated by adding the series elements:

$$R_T = R_1 + R_{2,3} = 250 + 66.67 = 316.67 \ \Omega$$

Now that you have the source voltage $E_T = 100$ V and the total resistance R_T, you can solve for the total current, I_T, using the following formula:

$$I_T = \frac{E_T}{R_T} = \frac{100}{316.67} = 0.316 \ \text{A}$$

You now have all three parameters for the equivalent circuit. With these values, you can find the total power for the circuit P_T and the power consumed by each of the circuit components R_1, R_2, and R_3.

First you solve for the total circuit power, P_T, using the following formula:

$$P_T = E_T \times I_T = 100 \times 0.316 = 31.6 \ \text{W}$$

Now you have the needed information to solve for the individual power dissipations starting with R_1. Since in a series circuit the current is the same throughout the circuit and you know the total current I_T, the current flow in R_1 and $R_{2,3}$ can be calculated as

$$I_T = I_1 = I_{2,3} = 0.316 \ \text{A}$$

Now you have the current and the resistance for R_1, so you can solve for the power consumed by R_1 using the following formula:

$$P_1 = I_1^2 \times R_1 = (0.316)^2 \times 250 = 0.099856 \times 250 = 24.964 \ \text{W}$$
$$\text{(which is rounded to 25 W)}$$

There are several ways that you can calculate the power dissipated by R_2 and R_3. Since you know the current through the parallel branch $R_{2,3}$ and the equivalent resistance of $R_{2,3}$ (calculated earlier in the problem), you can calculate the voltage across this branch using Ohm's law. Use the following formula:

$$E_{2,3} = I_{2,3} \times R_{2,3} = 0.316 \times 66.67 = 21.07 \ \text{V}$$

Since this is a parallel branch, you know that the voltage is the same across both R_2 and R_3:

$$E_{2,3} = E_2 = E_3 = 21.07 \ \text{V}$$

Now you have the voltage drop and resistance of R_2 and R_3 and can calculate the power dissipated by each of those components using the following formulas:

$$P_2 = \frac{E_2^2}{R_2} = \frac{(21.07)^2}{100} = 4.439 \ \text{W}$$
$$P_3 = \frac{E_3^2}{R_3} = \frac{(21.07)^2}{200} = 2.219 \ \text{W}$$

You have completed all the calculations required to find the total power and the power consumed by the individual components that made up the circuit in Figure 18–1. Now you can check your calculations by adding the individual power dissipations and comparing them to the total power calculated earlier:

$$P_T = P_1 + P_2 + P_3 = 24.964 + 4.439 + 2.219 = 31.622 \text{ W}$$

Earlier you calculated P_T as $0.316 \times 100 = 31.6$ W. The slight error is caused by rounding off during the calculations.

18.2 Circuits with Known Power Usage

The circuits that you will work on in your job will often have known values of power and voltage. One of the aspects that you will have to consider is the real power requirement for circuits with components that are already rated. For example, look at Figure 18–2. In this circuit, three lamps are connected to the power source using 300 feet of #16-AWG copper wire. You can see from the picture that each lamp is rated at 25 watts each and that the lamps are connected in parallel. When you look at a circuit like this one, you must be sure to take all the circuit parameters into account. The electrician who installed the lamps calculated that the lamps would draw ½ ampere each, making the total circuit current the sum of the current in the individual branches, or 1.5 A. The following formulas support the electrician's calculations.

To calculate the current of the individual lamps (the total current is the sum of all three lamps),

$$I_{total} = I_{lamp \#1} + I_{lamp \#2} + I_{lamp \#3} = 0.5 + 0.5 + 0.5 = 1.5 \text{ A}$$

This all seems to check out fine; however, look at the circuit again. The load, all three lamps, is 300 feet from the source voltage. You know from earlier chapters that all conductors, even copper, have some resistance. You also know that the longer the wire, the higher the resistance: #16-AWG copper has a resistance of 4.99 Ω per 1,000 feet. This information comes from the *National Electrical Code®*, chapter 9, Table 8, and applies to seven-stranded wire at 75°C. You will notice that this value differs slightly from the value given in Table 3–1. This

FIGURE 18–2   A real combination circuit.

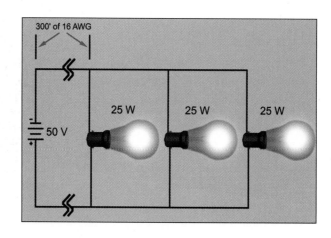

is because the value in that table is for solid conductor wire, not stranded wire.

Review other sizes of wire to see how much resistance they have per thousand feet of wire. Note that the larger the wire, the lower the resistance value. The fact that the load is 300 feet from the power source means that the resistance of the copper #16 conductor, both going to and returning from the lamps, must be considered in the calculations.

Since resistance exists in the conductors supplying the lighting load, the calculations for this circuit have other factors to consider. Look at Figure 18–3. This circuit was redrawn to represent all the resistances that must be considered in the circuit when performing the calculations.

The power rating of a lamp is the amount of power that the lamp will dissipate with rated voltage applied. Since rated voltage will not be applied because of the voltage drop across the wire, you need to know the lamp resistances so that you can solve this circuit as a combination circuit using the methods you have learned in this and previous chapters.

Since the lamps are rated at 25 watts at 50 volts, you can calculate the resistance value of the lamp as follows:

$$R_{lamp} = \frac{E_{lamp}^2}{P_{lamp}} = \frac{50^2}{25} = 100\ \Omega$$

Note that this is the resistance of each individual lamp.

To completely analyze this circuit you also need to know resistance of the wire feeding the lamps. Based on the resistive value of #16-AWG wire being 4.99 Ω per 1,000 feet of wire, you can calculate the total resistance as

$$R_{wire} = \frac{4.99\ \Omega}{1,000\ ft} \times 300\ ft = 1.497\ \Omega$$

Remember, this circuit has 300 feet of wire going to the lamps and 300 feet of wire coming back from the lamps. Both distances must be

FIGURE 18–3   Schematic diagram of lamp circuit (Figure 18–2) including resistance of the wires.

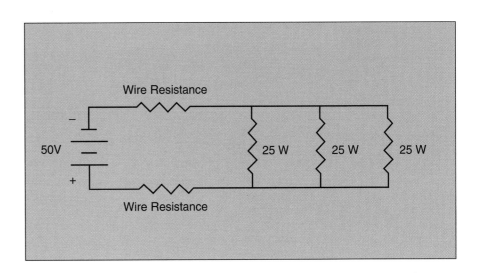

taken into account. Figure 18–4 shows the complete circuit with all resistances included.

This is a combination circuit that can be reduced using the methods you have learned. Start with the calculation of the resistance of the three lamps in parallel, $R_{\text{all lamps}}$.

$$R_{\text{all lamps}} = \frac{R_{1 \text{ lamp}}}{N} = \frac{100}{3} = 33.33 \ \Omega$$

Note that N is the number of lamps.

The circuit has now been reduced to a simple series circuit with the two lengths of copper wire and the parallel combination resistance of the lamps. The total resistance is the sum of the three series elements:

$$R_T = R_{\text{wire \#1}} + R_{\text{wire \#2}} + R_{\text{all lamps}} = 1.497 + 1.497 + 33.33 = 36.324 \ \Omega$$

The total current is calculated as

$$I_T = \frac{E_T}{R_T} = \frac{50}{36.324} = 1.38 \ \text{A}$$

Notice that the value that was calculated by the electrician installing this project (1.5 A) is different than the value calculated previously. The reduction in current is caused by the added resistance of the wire.

Now we will find out how much voltage is actually supplied to the lamps. We know the total current for this circuit is actually 1.38 A, and we know the equivalent resistance for the lamps that is in series with the resistance equals 33.33 Ω. Using Ohm's law, we can easily calculate the voltage across the lamps as follows:

$$E_{\text{lamps}} = I_{\text{lamps}} \times R_{\text{lamps}} = 1.38 \times 33.33 = 46 \ \text{V}$$

This means that we have lost, or "dropped," 4 volts on the wires. In percentage, this is

$$\% \ \text{Drop} = \frac{E_{\text{rated}} - E_{\text{actual}}}{E_{\text{rated}}} \times 100 = \frac{50 - 46}{50} \times 100 = 8\%$$

FIGURE 18–4   Schematic diagram of Figure 18–2 showing actual resistance values.

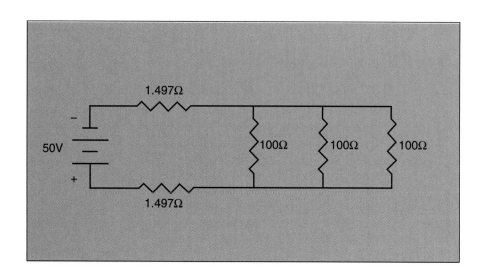

This much drop is excessive. Maximum voltage drops should normally be limited to no more than 3%.

The lamps are not dissipating the amount of power for which they were designed. In other words, instead of the lamps actually providing light based on their design parameters, the lights will now be operating based on the actual installation parameters. Let's see what effect the installation parameters actually have on the lamps.

Since we now know that the installation is being supplied by 46 volts to the lamps, we can recalculate the actual wattage of the lamps as follows:

$$P_{lamp} = \frac{E_{lamp}^2}{R_{lamp}} = \frac{(46)^2}{100} = 21.16 \text{ W}$$

This circuit, by taking into account the resistance of the wires, in reality has lost approximately 15% of the lamp output. This is

$$\left(\% \text{ Loss} = \frac{P_{rated} - P_{actual}}{P_{rated}} \times 100 = \frac{25 - 21.16}{25} \times 100 = 15.4\% \right)$$

You can see that the whole circuit must be taken into account when designing an installation. All the resistance components in a combination circuit consume power and must be taken into account. Take a look at the resistance of the wires and see how much power their resistance consumes.

Since you have already calculated the current and the resistance of the wires, you can easily calculate the power consumed by the wires from the following formula:

$$P_{wires} = 2 \times (1.38^2 \times 1.497) = 5.7 \text{ W}$$

Note the multiplier of 2. There are two wires, and you must consider both of them.

What the previous calculation tells you is that out of all the power consumed by the installation, a portion of that power, 5.69 watts, is consumed by the wires and wasted as heat. By not taking into account the power lost in the wires, the actual circuit did not perform to optimum design efficiency, the actual lamp output was reduced, and the wires consumed power that was wasted as heat. All of this may have been avoided or reduced by selecting a larger wire size.

■ SUMMARY

The power in all combination circuits is consumed by all the resistive components in that circuit. The total power consumed in any circuit is the sum of the power consumed by each individual component.

When designing or working on a combination circuit, all the resistive components, including the wires that supply the components of the circuit, must be taken into account.

■ REVIEW QUESTIONS

1. State and discuss the formulas for solving power.
2. What effect does the length of the circuit conductors have when calculating power or voltage drop? How can this effect be modified or improved?

3. Where do you find the resistance rating for wire?
4. What is the maximum voltage drop allowed on a branch circuit? What can you do to improve voltage drop?

■ PRACTICE PROBLEM

1. A commercial warehouse is 1,000 feet long. Each end of the warehouse has two 400-watt, 120-volt HPS wall pack-type light fixtures mounted on it. The supply panel is a 120-volt panel and is located in the middle of the warehouse. The wall packs are fed with #12 stranded copper wire.

   a. Draw the equipment diagram and the schematic diagram.

   b. What is the total current of the circuit?
   c. What is the power consumed by each element of the circuit?
   d. What is the total power consumed by the circuit?

PART

5

MAGNETISM AND GENERATORS

CHAPTER 19

Understanding the Principles of
Magnetism and Electromagnetism

CHAPTER 20

How Electrical Generators Work

Understanding the Principles of Magnetism and Electromagnetism

■ OUTLINE

■ OVERVIEW

Magnetism is a force that occurs in nature. It is most often expressed as the ability of some materials to attract iron or other ferromagnetic materials. Most electricity is developed through the use of magnetism. No one knows exactly when magnets were first discovered, but the Greeks were studying magnetism and magnets over 2,000 years ago. The Chinese may have been studying magnetism even before then; however, most Western terminology is based on the Greek studies. They discovered that iron was attracted to a certain kind of stone. This stone was originally discovered in a region of Asia called "Magnesia" and was called "magnetite." From this word comes our modern word "magnet."

■ OBJECTIVES

After completing this chapter, you should be able to:

1. Demonstrate your understanding of magnetic materials through classroom discussion.
2. Explain the theory of magnetism.
3. Describe how electron flow creates a magnetic field and how that field reacts to physical changes.
4. Explain how electromagnetism performs useful and meaningful work.

■ GLOSSARY

Ampere-turn The strength of an electromagnetic field calculated by multiplying the current flow times the number of turns.

Ferromagnetic Properties displayed by certain substances, such as iron, nickel, or cobalt and various alloys, that exhibit extremely high magnetic capability.

Magnetic flux The force of a magnetic field expressed as lines. Also called magnetic lines of force.

Magnetic lines of force See Magnetic flux.

Magnetite The mineral form of black iron oxide. Magnetite is a naturally occurring magnet and was possibly the first type of magnet studied by the ancients.

Permeability The degree to which a material focuses magnetic lines of flux.

Reluctance The opposition that materials present to the flow of magnetic lines of flux.

Saturation The point at which the magnetic domains are all lined up. Any additional lines of force will not flow through the core but will leak around the core.

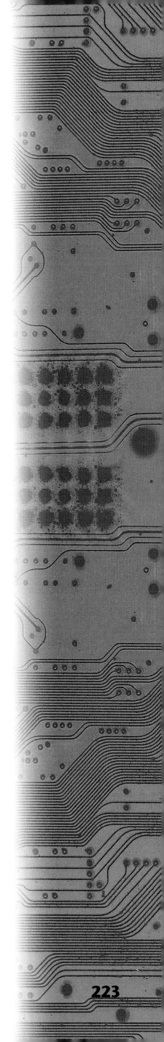

■ THE NATURE OF MAGNETISM

19.1 Natural Magnets

Magnets were first discovered as being useful when magnetite was observed to consistently align itself to the same orientation when floating in water. Ancient Greeks placed magnetite on a piece of wood floating in water. Regardless of the magnetite's initial placement, it would eventually return to the same position. This phenomenon occurs because Earth is also a magnet. Modern scientists believe that the motion of Earth's molten iron core generates a huge magnetic field as shown in Figure 19–1.

19.2 Magnetic Reactions

If two magnets are held close to each other, they will either attract or repel. The attraction or repulsion is caused by the magnetic lines of force that flow into and out of a magnet. Look at Figure 19–2. Notice that the lines of force flow out of the north end of the magnet and flow into the south end of the magnet.

When the ancient Greeks first learned the one end of a magnet would always point north, they decided to call that end of the magnet the "north" pole. Later researchers learned that like poles attract and that unlike poles repel. Clearly, the magnet's north pole and Earth's magnetic pole must be opposite. Because of this, a magnet pole that is attracted north is called the "north-seeking" pole. Usually, we just shorten this to the magnet's north pole.

A naturally occurring magnet *always* has a field around it, so we call it a *permanent magnet*. Permanent magnets display one of the laws of magnetism. This law states that energy is required to build a magnetic field but that no energy is required to maintain one. With mod-

FIGURE 19–1   Earth as a magnet.

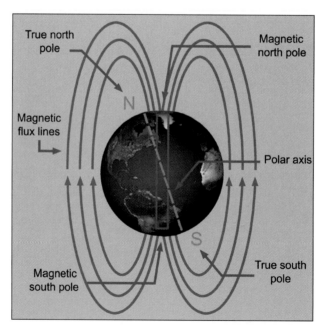

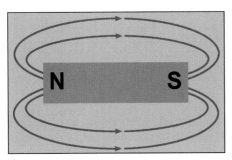

FIGURE 19–2   Magnetic lines of force.

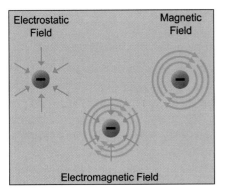

FIGURE 19–3   Individual electron fields.

ern technology, very strong permanent magnets can be manufactured. These magnets are stronger and last longer than naturally occurring magnets.

19.3 Theory of Magnetism

Through intense scientific study of magnets and the materials from which natural magnets are made, a theory of magnetism has been developed. This theory is based on the ability of a material to be magnetic because of the "spin" of the electrons in the material.

In chapter 1, you learned that electrons orbit atoms. Being a charged particle, electrons have an electrostatic field around them. The lines of force from this field travel toward the electron (see Figure 19–3). Electrons also spin on an axis, much like Earth. This makes each electron act like a small electromagnet. The combination of the electrostatic and magnetic fields creates a composite electromagnetic field somewhat like the one shown in Figure 19–3. Notice that the magnetic lines and electrostatic lines are perpendicular to each other.

All electrons do not spin in the same direction. Electrons that are in the same shell and spin in opposite directions usually form pairs. Because of their opposite spin, the magnetic properties of these paired electrons usually cancel each other. This is why all but a few materials do not exhibit natural magnetism.

The iron atom has four unpaired electrons in its outer shell. These electrons are spinning in the same direction, causing their individual magnetic fields to add. These magnetic fields develop in small packets called *magnetic domains,* which react like small permanent magnets.

Most materials that have magnetic domains are not magnetized because the domains are not arranged properly. The domains do not lie in the same direction (Figure 19–4).

In magnetic materials, many or most of the domains are aligned. This alignment allows the magnetic properties of the domains to build, making all the small permanent magnets into one large permanent magnet (see Figure 19–5). The more the domains line up, the stronger the magnet becomes.

FIGURE 19–4   Magnetic domains not in order.

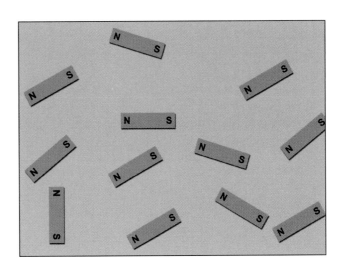

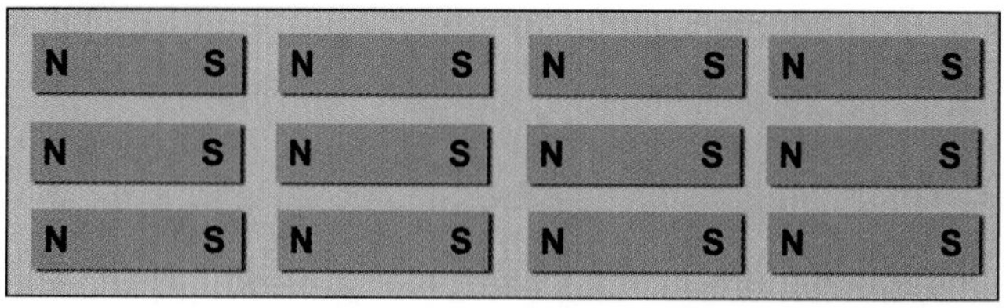

FIGURE 19–5   Magnetic domains aligned to create a magnetic field.

■ MAGNETIC FLUX

The lines of force that leave the north end of a magnet and enter the south end are called **magnetic flux**. You can see the effects of magnetic flux by placing a magnet under a piece of paper and sprinkling iron shavings onto the paper. The filings will take the shape of the field lines similar to those shown in Figure 19–2. Notice that the lines of flux stay parallel to each other and never cross. The same numbers of lines that leave the north pole enter the south pole. They are the densest (closest together) near the poles and the least dense when farthest away from the poles. Each line of flux has the same energy. The closer the lines of flux are to each other, the stronger the resulting magnetic field. Therefore, the strongest field is near the poles.

Lines of flux leave the north end of the magnet and enter the south end. In Figure 19–6, the top magnets have their north poles facing each other. They repel because the lines of flux are moving in opposite directions and pushing against each other. The bottom pair attracts because the lines of flux move in the same direction and aid each other.

19.4 Types of Magnetic Materials

Materials are classified into three categories: ferromagnetic, paramagnetic, and diamagnetic. **Ferromagnetic** materials include cobalt, nickel, iron, and manganese and can be easily magnetized. Paramagnetic materials include chromium, platinum, and titanium and can be magnetized but not as easily as ferromagnetic materials. Diamagnetic materials are actually slightly repelled by magnetic fields. Two of the strongest diamagnetic materials are graphite and bismuth.

Magnetic lines of flux will pass through all materials to one degree or another. Ferromagnetic materials allow the lines of magnetic flux to pass through easily and actually tend to focus the lines of flux. The ability of a material to focus lines of flux is called **permeability**. The opposition to magnetic lines of flux is called the **reluctance**.

FIGURE 19–6   Magnetic lines of flux repelling (top) and attracting (bottom).

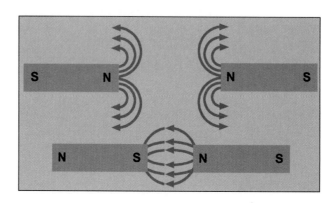

The best permanent magnets are made of steel, which is an alloy of iron and other metals. A magnet made solely of soft iron does not maintain its magnetism long. For this reason, equipment that requires a strong permanent magnet will have its magnet made of a steel alloy. Those pieces of equipment that need to be magnetized but should not retain their magnetism (like motors and generators) will have their structures made of soft iron.

■ ELECTRICITY AND MAGNETISM

19.5 Electromagnets

Applications

There are some applications in which a magnet is useful but only part of the time. For example, when you press a doorbell button, a small magnet pulls on a piece of ferrous metal, causing it to move and strike the chime, giving us the bell sound. This magnet does its work only when the button is pressed. Permanent magnets cannot do this. Permanent magnets cannot be turned off and on, so a different type of magnet, one that can be turned off and on, must be used: an electromagnet.

Polarity

When current flows through a wire, a magnetic field is produced in a concentric circular form around the wire (see Figure 19–7). The two circles are cross-sectional views of wires, and both have current flowing through them. The one on the left has current flowing "out" of the page. The right one has current flowing "into" the page. The arrows on the concentric circles show the direction the magnetic fields move. The fields around the wire on the left move clockwise, and the fields around the wire on the right move counterclockwise. The wire at the bottom shows the direction of the magnetic field with the current flowing from left to right.

FIGURE 19–7 Polarity of an electromagnetic field.

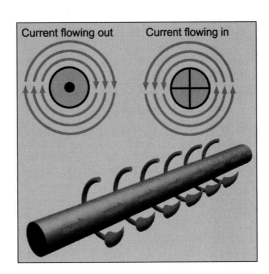

Current flowing out   Current flowing in

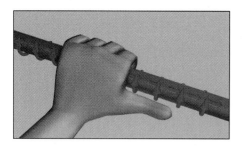

FIGURE 19–8   Left-hand rule for conductors.

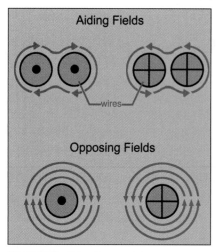

FIGURE 19–9   Field interactions in wires.

The direction that the field encircles the conductor can be determined by using a simple memory device called the left-hand rule for conductors (see Figure 19–8). This rule states that if you wrap your left hand around the conductor, with your thumb pointing in the directions of current flow, your fingers will encircle the conductor in the direction of the magnetic field surrounding that conductor.

Attraction and Repulsion

You learned in previous chapters that the definition of the ampere is derived from the amount of force between two parallel wires. This force comes from the interaction of the magnetic fields coming from the wires while current flows through them. Figure 19–9 shows three sets of parallel wires. The top two sets show current flowing in the same direction. In these two cases, the magnetic fields aid each other because at their points of interaction the fields are flowing in the same direction. This causes the wires to be attracted to each other. The interactions of the fields in the bottom pair of wires oppose each other. This opposition repels the wires.

Strength

The strength of an electromagnetic field can be increased by coiling the wires so that the magnetic field from each wire adds to the one next to it (see Figure 19–10). This aiding current creates one large magnetic field, with a north and south pole, like a permanent magnet. These coils are called *electromagnets*.

The strength of the field depends on several factors, including the number of coils, the current magnitude, and the spacing between the coils. More coils create a larger magnetic field. Higher current flow creates a larger field around the wire and thus a larger overall field. Multiplying the number of turns times the current flow results in a measurement called **ampere-turns**. Ampere-turn is the relative field strength of the coil. Wrapping the coils closer together, either by making the circles smaller and reducing the cross-sectional area or by reducing the space between coils, further increases the strength of the field because the amount of loss is decreased (see Figure 19–11).

FIGURE 19–10   Magnetic field produced in a coil of wire.

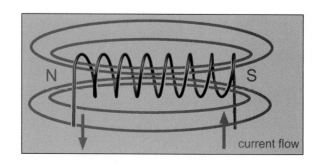

FIGURE 19–11 Relative magnetic strength of different coils.

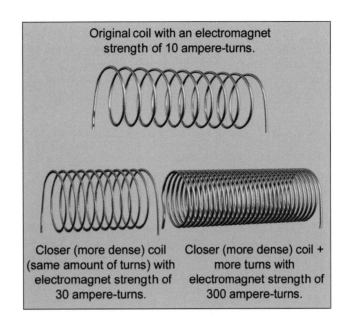

Original coil with an electromagnet strength of 10 ampere-turns.

Closer (more dense) coil (same amount of turns) with electromagnet strength of 30 ampere-turns.

Closer (more dense) coil + more turns with electromagnet strength of 300 ampere-turns.

Electromagnet Cores and Saturation

One way to create a stronger electromagnet is to increase the number of flux lines that pass through the coil's center. You learned earlier that certain types of materials (ferromagnetic and paramagnetic) cause the lines of flux to focus. These lines of magnetic flux focus because they tend to pass through the ferromagnetic and paramagnetic materials better than through air. If a coil were wrapped around a core of ferromagnetic material, the number of flux lines passing through the center of the coil would be increased and so would the strength of the electromagnet. This happens because the individual molecules in the core material become polarized (aligned) in relation to the magnetic field produced by the coil.

If the magnetic field becomes too strong (too much current or too many coils), the magnetic core will saturate. The **saturation** point is the point at which increasing the amount of current has no more effect on the magnetic field strength. This occurs because all the molecules that can be aligned magnetically have been. If the electromagnet needs to be stronger, then a different core material has to be chosen (see Figure 19–12).

A good material for a core is soft steel or iron. Soft iron has a high permeability—it easily passes magnetic lines of flux and makes the electromagnet stronger. Furthermore, soft iron is a good core because it does not become permanently magnetized.

Electromagnet Polarity

To determine the orientation of the north pole of an electromagnet, use the left-hand rule (see Figure 19–13). To use the left-hand rule, wrap the fingers around the coil pointing in the direction of current flow. In this position, the thumb points in the direction of the north pole of the electromagnet.

FIGURE 19–12   Electromagnetic cores.

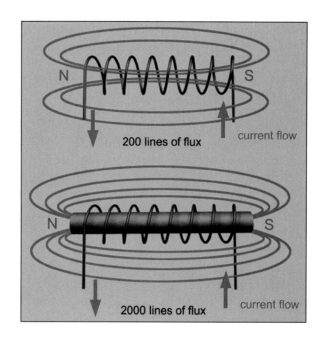

FIGURE 19–13   Left-hand rule for determining magnetic polarity.

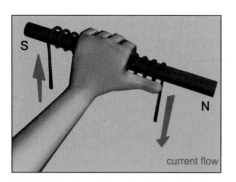

FIGURE 19–14   Induced voltage causing current flow.

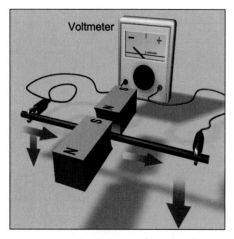

19.6 Inducing Voltage with Magnetism

All of the previous discussion has centered on the magnetism and magnetic field. A magnetic field is generated around a conductor that is carrying current. This section discusses the reverse principle, which says that a voltage will be induced into a conductor that is passed through a magnetic field. If this conductor is connected to a complete path, current will flow. Note that in Figure 19–14 a complete circuit is not shown. Obviously, there must be a complete circuit for current to flow.

To better understand this phenomenon, refer to Figure 19–14. The conductor is being moved downward between the poles of the two magnets. The poles of the magnets create a field that is flowing from north pole to south pole. This downward motion of the conductor causes a voltage to be induced into the conductor, the polarity of which causes the current to flow in the direction shown if a complete circuit exists. The electrons move because of the interactions between the magnetic field of the permanent magnet and the electromagnetic fields of the electrons.

If the motion is upward, the magnetic interactions cause the voltage polarity to reverse the flow of current as shown in Figure 19–15. This is because the direction of the field interactions is opposite. The same would be true if the poles of the magnets were reversed and the motion were downward. The field interactions would be reversed, and current would flow in the opposite direction.

A rule that can be used to determine the direction of current flow is the left-hand rule for generators (see Figure 19–16). This rule says that with the thumb, index, and middle fingers of the left hand at right angles to one another and the thumb pointing in the direction of motion and index finger pointing in the direction of the magnetic field flux (north to south), the middle finger points in the direction of current flow.

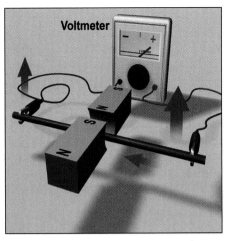

FIGURE 19–15   Motion in the opposite direction causes the current to change direction.

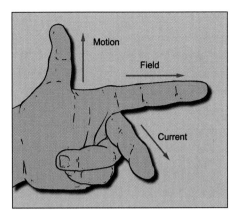

FIGURE 19–16   Left-hand rule for generators.

An important point is that for the current to flow, there must be relative motion between the conductor and the magnetic field—either the conductor must move through the field or the field must move around the conductor. Notice also that the conductor and the field flux are perpendicular to one another (see Figures 19–14 and 19–15). This is necessary for the conductor to "cut" the lines of flux. If no lines of flux are cut, then no voltage will be induced. If the conductor were parallel to the lines of flux, even if there were relative motions, the conductor would cut no flux lines, and no voltage would be induced.

■ SUMMARY

Magnetism is the ability of a material to attract iron or other ferromagnetic materials, such as nickel, cobalt, and manganese. Materials that have this ability are referred to as magnets. Magnets have two magnetic poles: a north-seeking pole and a south-seeking pole. If a magnet is suspended and free to spin, by definition the north-seeking pole of the magnet is the pole that points toward the magnetic north pole of Earth (opposites attract). Note that even though the north-seeking pole is actually opposite to Earth's north pole, it is still referred to as the north pole. Like poles of magnets repel each other; unlike poles attract each other.

Magnets are generally divided into two groups: permanent magnets and temporary magnets. Permanent magnets, once they become magnetized, retain their magnetism indefinitely. They usually have cores made of steel. Temporary magnets lose their magnetic properties when their magnetizing force is removed. Their cores are usually made of soft iron.

Magnets have lines of force called flux lines. Flux lines are invisible to the human eye, but their presence can be detected by sprinkling iron filings on a piece of plastic that is placed on top of a magnet. Flux lines have some specific rules and characteristics that must be understood in order to study magnetic-electrical principles:

1. They flow from the north pole of a magnet to the south pole.
2. They always form closed loops. The number of flux lines leaving the north pole of a magnet must eventually return to its south pole.
3. They never cross each other.
4. They tend to cancel each other when they meet flowing in opposite directions.
5. They repel each other as they leave the magnet's north pole.
6. They take the path of least reluctance.

7. They pass through all materials.

8. They are denser at the poles of the magnet. As the flux lines move away from the north pole of the magnet, they repel each other and become less dense.

9. They always try to find the shortest, lowest reluctance path from the north pole to the south pole.

Whenever a conductor cuts through a magnetic field, a voltage will be induced into that conductor. If the conductor is part of a circuit, current (electrons) will flow. Reversing the direction that the conductor travels through the magnetic field will result in change in the direction of the electron (current) flow. The direction of the current can be determined by using the left-hand rule for generators as shown in Figure 19–16.

When current flows through a straight conductor, a magnetic field is created. This field can be easily detected by passing a compass over a current-carrying conductor and noting how the needle is deflected. When the current is turned off, the conductor's magnetic field collapses, and the north compass needle will once again point to Earth's magnetic north pole.

The magnetic field around a conductor is a series of concentric flux lines situated perpendicular to the conductor. They do not form a magnetic polarity. The direction that the field encircles the conductor can be determined by using the left-hand rule for conductors. The left-hand rule states that if you wrap the left hand around the conductor, with the thumb pointing in the direction of the current flow, the fingers will encircle the conductor in the direc-

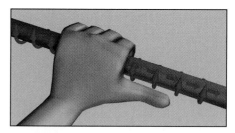

FIGURE 19–17   Left-hand rule for conductors.

tion of the magnetic field surrounding that conductor (see Figure 19–17).

If the conductor is wound in the shape of a coil, a magnetic field (with polarity orientation) will be formed when current flows through it. The polarity of the magnetic field can be determined by using the left-hand rule for coils. If you encircle the coil with your left hand, placing your fingers so that they point in the direction of the current flow, then your thumb will point toward the north magnetic pole. Refer back to Figure 19–13.

The strength of the magnetic field of a coil depends on several factors:

1. The amount of current

2. Number of turns

3. Core permeability

4. Cross-sectional area of core material

5. Length of core and spacing between turns

6. How windings are wrapped (e.g., on top of each other or woven together). In general, the greater the frequency of adjacent turns in closer proximity, the greater the inductance.

■ REVIEW QUESTIONS

1. Define magnetism.
2. Name some ferromagnetic materials.
3. Discuss how magnetic poles repel and attract.
4. Magnets are divided into two groups. What are they?
5. Which type of magnet is an electromagnet?
6. Which direction do the lines of flux flow in a magnet?

7. Which material is better for a permanent magnet: steel or soft iron?
8. Give some examples of electromagnets.
9. What type of core is used in an electromagnet?
10. Do flux lines ever cross?
11. Name some of the factors that determine the strength of the magnetic field of a coil.

chapter 20

How Electrical Generators Work

■ OUTLINE

OVERVIEW

English physicist Michael Faraday discovered the basic theory of electric generation in 1831. He discovered that a magnetic field could be used to produce an electrical current. Practically all commercial power is produced using the principle of magnetism.

Mechanical energy delivered from raw manpower, water, wind, fossil fuels (coal and oil), and nuclear reactors is used to turn the rotors of a generator. The energy from some of these sources, such as fossil fuels and nuclear reactors, must first be converted into another form of energy in order to turn the rotor of a generator. The topics in this chapter are based on the fundamentals of electrical generation and are intended to show the relationship between magnetism (which you have previously studied) and the production of electrical power.

Note that in this chapter you will learn only the basics of electrical generators. Later in your education, you will spend a great deal of time learning the details of "real world" rotating equipment.

OBJECTIVES

After completing this chapter, you should be able to:

1. Explain the basic principles involved in generating electromotive force.
2. Describe the basic construction and operation of simple DC and AC generators.
3. Demonstrate the left-hand rule for generators to show relative motion, magnetic flux, and electron flow.

GLOSSARY

Electromagnetic induction The generation of electricity by passing a conductor through a magnetic field. The same thing can be accomplished by moving a magnetic field across a conductor.

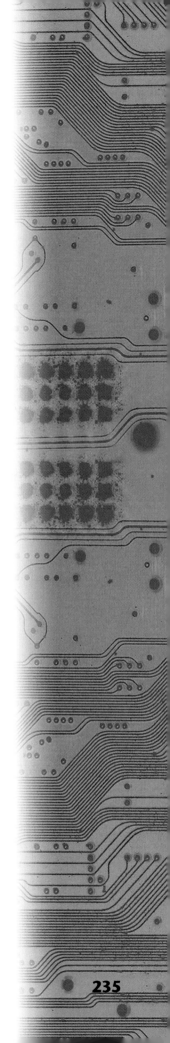

GENERATING ELECTRICITY USING MAGNETISM

20.1 Electromagnetic Induction

Earlier you studied the concept of **electromagnetic induction**. You learned that a voltage is induced into a conductor that is passed through a magnetic field. If this conductor is connected to a complete path, or closed loop, current will flow.

Figure 20–1a shows a conductor being moved downward between the poles of the two magnets. The poles of the magnets create a field that is moving from the north to the south pole. The downward motion of the conductor causes a voltage to be induced into the conductor, the polarity of which causes the current to flow in the direction shown. The electrons move because of the interactions between the magnetic field of the permanent magnets and the electromagnetic fields of the electrons themselves. Note that without movement, there is no current flow.

If the motion is upward, the magnetic interactions cause the voltage polarity to reverse the flow of current as shown in Figure 20–1b. This is because the direction of the field interactions is opposite. The same would be true if the poles of the magnets were reversed and the motion were downward. The field interactions would be reversed, and current would flow in the opposite direction. One of the key factors to remember is that relative motion between the magnetic field and the conductor must exist before a voltage will be generated. This motion can be the motion of the magnetic field instead of the conductor; that is, the magnets can move instead of the conductor.

The left-hand rule for generators (see Figure 20–2) says that with the thumb, index, and middle fingers of the left hand at right angles to one another and the thumb pointing in the direction of motion and index finger pointing in the direction of the magnetic field flux, the middle finger points in the direction of current flow.

FIGURE 20–1 The polarity of the voltage created by a magnetic field depends on the direction of motion: (a) downward motion; (b) upward motion.

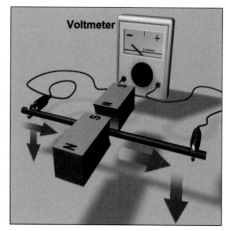
(a)

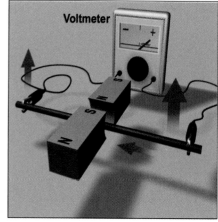
(b)

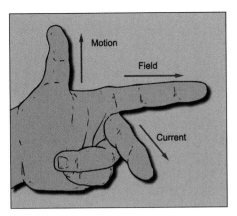

FIGURE 20–2   Left-hand generator rule.

20.2 Calculating the Amount of Induced Voltage

The magnitude of the voltage being produced by electromagnetic induction is determined by how many lines of magnetic flux are being cut by the conductor per second. For the simple system such as that shown in Figure 20–1, three main factors determine this magnitude: the strength of the magnetic field, the speed of the wire cutting the magnetic field, and the number of loops in the wire. To produce 1 volt, the wire must cut through 100,000,000 lines of flux in 1 second. Field strength equal to 100,000,000 lines is called a *weber* (Wb). If a wire moved through twice that number of flux lines in 1 second, 2 volts would be produced. Also, if the same wire cut through 1 Wb in half the time, 2 volts would be produced. If the number of loops in the wire is increased, it effectively increases the number of flux lines being cut per second so that a larger voltage is produced.

As the loop wire in Figure 20–3 rotates through one complete turn, it will generate voltage. Figure 20–4 shows the waveform of the voltage being generated.

■ GENERATOR CONSTRUCTION AND OPERATION

20.3 AC Generators

The understanding of a DC generator is best started by learning about an AC generator. Look at Figure 20–3. This simple AC generator is constructed of a permanent magnet, slip rings, brushes, a single loop (armature), and a voltmeter (galvanometer) to read generated voltage. The armature is the rotating part of the machine in which the voltage is induced, and the magnets set up the required magnetic field.

Each end of the wire loop is connected to a slip ring. A slip ring is a component that allows the current that is induced into the rotating coil to be connected to an external (nonrotating) load. The slip ring

FIGURE 20–3   Simple AC voltage generator.

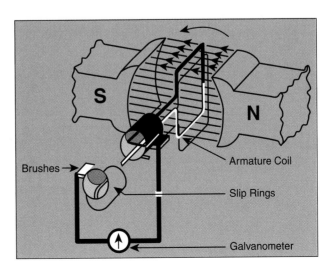

FIGURE 20–4 One cycle of induced voltage.

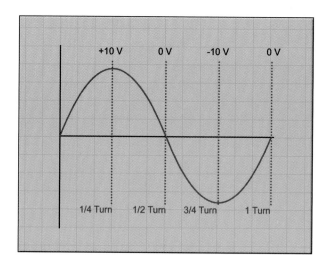

The opposite sides of the loops are moving in opposite directions when compared to each other. The left side of the generator is moving up while the right side is moving down. This means that the current comes out of one side of the loop for half a revolution and comes out of the other side for the other half revolution.

mechanism consists of a conductive ring (generally copper or brass) that is connected to the armature coil. A second component, a brush, makes contact with the slip ring, allowing it to turn but still providing electrical continuity. Brushes are normally made out of carbon or graphite with a wire conductor tail that can be attached to the external circuit or load. The brushes are spring loaded to help maintain the correct brush pressure on the slip ring. Two brushes are required for a simple single-phase AC generator, one for each end of the armature loop. As the loop rotates, each side of the loop will be cutting lines of flux at the same time. Figure 20–5 is a drawing of a brush and brush holder.

The opposite sides of the loops are moving in opposite directions when compared to each other. The left side of the generator is moving up while the right side is moving down. This means that the current comes out of one side of the loop for half a revolution and comes out of the other side for the other half revolution.

FIGURE 20–5 Brush and brush holder.

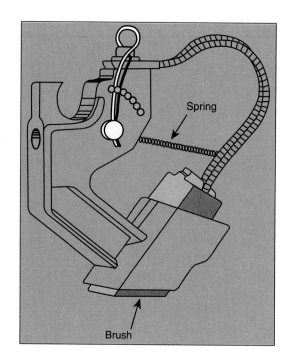

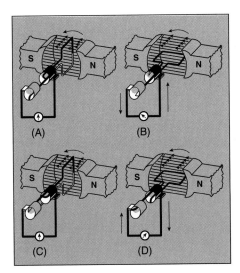

FIGURE 20–6   Generation of an alternating voltage.

Figure 20–4 shows a waveform of the output of the simple AC generator as the armature loop rotates through the magnetic field. Notice the change in that waveform for the various positions of the loop relative to the magnetic field.

In Figure 20–6a, the loop is rotating but is cutting very few lines of flux. This position is called the 0° position. If the voltage were being measured here, it would be zero as shown on the galvonometer.

In Figure 20–6b, the generator has rotated 90°, and both sides of the loop are cutting through the most lines of flux because they are cutting them perpendicularly. At this point the generator is producing the highest voltage. If the voltage were being measured, it would show the output of the generator to be at its highest positive value—10 volts as indicated in Figure 20–4.

In Figure 20–6c, the generator is producing the least amount of voltage again, but this time the loop is in the 180° position, opposite the position shown in Figure 20–6a.

Figure 20–6d shows the loop in the 270° position. It is now cutting the maximum number of flux lines and producing the maximum voltage. The voltage is now negative when compared to the voltage being produced in Figure 20–6b. This is because the current is now in the opposite direction through the loop.

Figure 20–7 shows the rotating loop and compares its positions to the output voltage as shown in Figure 20–4.

20.4 DC Generators

DC generators are similar to AC generators except that the voltage is removed from the armature by a device called a *commutator* instead of slip rings. A commutator is similar to a slip ring that has been cut into pieces or segments.

Now look at Figure 20–8. Each commutator segment is attached to the end of one side of each loop in the generator. Figure 20–9 shows

FIGURE 20–7   An alternating voltage sine wave.

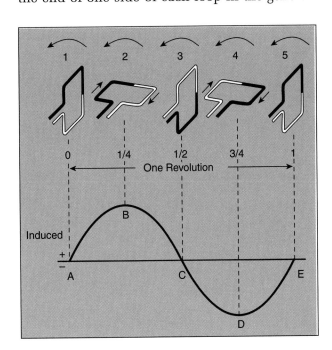

FIGURE 20–8 A simple DC
generator.

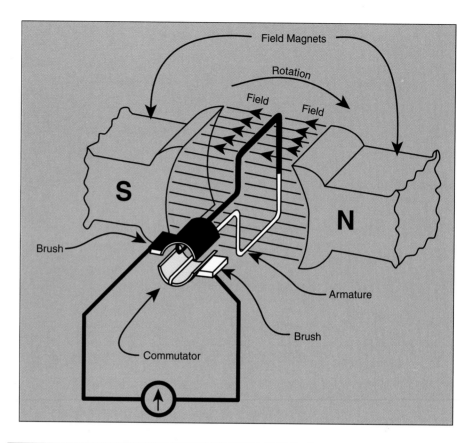

FIGURE 20–9 One revolution of
output from the generator shown in
Figure 20–8.

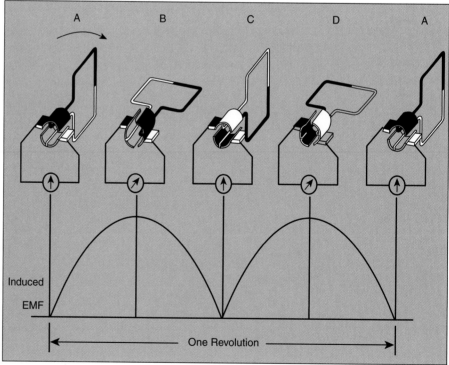

the output voltage at the brushes. For the first half a rotation, the output of the DC generator is exactly the same as the AC generator. However, when the armature loop reaches the 180° point in its rotation, the commutator segments switch brushes so that the output to each brush maintains the same polarity.

20.5 Parts of a Generator

The names of the parts of generators and motors are among the most often misunderstood of all electrical terms. Understanding these terms will help you as you progress through your career.

Stator and Rotor

The stator makes up all the nonrotating electrical parts of a generator or motor, and the rotor makes up all the rotating parts. This is true whether the machine is AC, DC, generator, or motor.

Field and Armature

The field of a machine is the part that generates the direct magnetic field. The current in the field does not alternate. In Figures 20–3 and 20–8, for example, the field is the permanent magnet and its associated lines of magnetic flux.

The armature winding is that which generates or has an alternating voltage applied to it or which generates an alternating current. Usually, the terms *armature* and *field* are applied only to AC generators, synchronous motors, DC motors, and DC generators. The following identifies which is which for each one:

AC Generators   The field of an AC synchronous generator is the winding to which the DC excitation current is applied. The armature is the winding to which the load is connected. In small generators, the field windings are often on the stator and the armature windings on the rotor. Most large machines, however, have a rotating field and a stationary armature.

AC Synchronous Motors   A synchronous motor is virtually identical to a large synchronous generator. Thus, the armature is the stator, and the field is the rotor.

AC Induction Motors and Generators   The terms *armature* and *field* are not normally applied to induction motors. The stator of the induction machine is identical to the stator of the synchronous motor and generator; consequently, the stator is the most closely associated with the term *armature*. The field is established in the rotor; however, the terms are still rarely used.

DC Motors and Generators   In DC motors and generators, the armature is the rotor, and the field is the stator. Because the armature is always the rotor on DC machines, many electricians and engineers mistakenly believe that the armature is the rotor on all motors and generators.

Slip Rings

In some AC machines, the rotor windings are connected to the outside world through slip rings (also called *collector rings*). As previously described, brushes ride on the slip rings and make an electrical connection between the rotor and the external circuit.

FIGURE 20-10   Commutator construction.

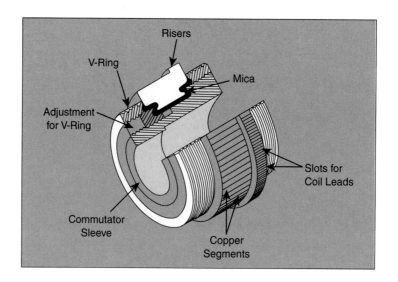

Commutator

In DC machines, a commutator is used instead of slip rings. The commutator has segments, with one pair for each loop in the armature. Each of the commutator segments is insulated from the others by an insulator material, such as mica. As discussed earlier, the commutator converts the AC machine to DC operation. In a generator, the commutator segments reverse the loop connections to the brushes every half cycle to maintain constant polarity of voltage. For a DC motor, the commutator segments maintain consistent polarity and allow the motor to produce torque in only one direction.

Earlier you saw a two-segment commutator. Figure 20–10 is a drawing of a multisegment commutator that is typical of those used in commercial machines.

Brushes

These are usually made of graphite and are spring loaded so the brush is held firmly against the spinning rings on the rotor or commutator. Brush leads are used to connect the armature to the external circuit. Figure 20–5 is a drawing of a single-brush assembly.

Pole Pieces

The magnetic field in most DC generators and motors is created using a DC electric current rather than the permanent magnets discussed earlier. If the field assembly is a permanent magnet, it will be made of a material that retains its magnetic field, such as steel. Electromagnetic field poles are made of laminated soft iron.

Pole pieces are found inside the generator housing. Figure 20–11 shows the frame of a DC motor with the two main pole cores and two smaller ones called *interpoles*. Figure 20–12 shows a pole piece with the wiring.

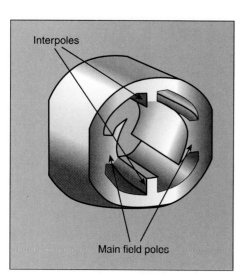

FIGURE 20-11   DC generator frame showing the poles.

FIGURE 20-12   Wound field pole.

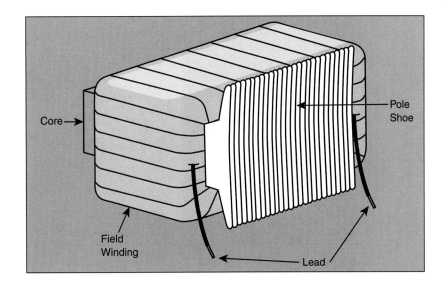

LENZ'S LAW

A coil of wire wrapped around a core is called an *inductor*. A voltage can be induced into a conductor by moving it through a magnetic field or moving a magnetic field through the conductor. This induced voltage produces a current. You also learned that passing current through a wire produces a magnetic field around it.

Heinrich Lenz discovered that the magnetic field created by the current moving in a conductor as a result of passing the conductor through a magnetic field opposes the magnetic field that caused the current to flow in the first place. This is the basic principle used to determine the direction of an induced voltage or current. Lenz's law states that an induced voltage or current produces a magnetic field that opposes the magnetic field that induced the voltage or current.

What this says about inductors is that if you change the current through an inductor, the magnetic field around the inductor will move (get either bigger or smaller). This movement of the field will cut the coils of the inductor, inducing a voltage into the wire. This induced voltage from the moving magnetic field will oppose the original change in current. Inductors always oppose a change in current.

SUMMARY

Most modern commercial electric power is produced by electromagnetic induction. In some AC generators and all DC generators, a rotating winding (the armature) passes through a magnetic field to induce voltage into itself. This is called a rotating armature.

In larger AC generators, the rotor produces a rotating magnetic field that cuts through the stator windings to produce output voltage. This second type of generator is called a rotating field.

Most generators in use today are AC, though there are still some applications for DC. AC generators with rotating armatures use slip rings to remove the voltage; DC generators use commutator segments. AC generators with rotating fields use the slip rings to apply the DC current to the field winding as it rotates.

A wire loop must pass through 1 Wb of flux in 1 second to produce 1 volt. The direction of current flow through the armature can be deduced using the left-hand rule for generators.

■ REVIEW QUESTIONS

1. By what principle is most electrical power produced today?

2. What are some of the energy sources that are used to power generators?

3. Electromagnetic induction causes a _____ to be induced into a _____ that is passed through a magnetic field or when the magnetic field is passed over the conductor.

4. In order for a voltage to be generated, you must have movement of what?

5. What are three factors that determine the amount of voltage induced in a generator?

6. Where are slip rings used on an AC generator? On a DC generator?

7. What is the most prevalent generator used today?

8. How is voltage removed from the armature of a DC generator?

9. Discuss the left-hand generator rule and how it affects AC and DC generators.

10. State Lenz's law.

PART

6

DC CIRCUIT ANALYSIS TOOLS

chapter 21

Understanding Voltage Polarity and Voltage Drop

■ **OUTLINE**

■ OVERVIEW

When you use a battery or a DC instrument such as a voltmeter or an ammeter, you will note that the terminals are marked with a "+" and a "−" sign. These symbols indicate the polarity of the terminals, and they must be understood and respected. Polarity is the property of a component that has two points with opposite characteristics. For example, a battery has two terminals (plus and minus), and a potential exists between those two terminals. The measure of that potential difference is called *voltage.*

Connecting some types of electrical or electronic equipment with their indicated polarities reversed can result in damage to or destruction of the device being used. For this reason, equipment of this type will normally have polarity marks indicating which terminal is positive or negative. Always check before you actually conduct a measurement with such equipment.

CAUTION: Failure to observe polarity during measurements may result in equipment damage and/or severe injury.

■ OBJECTIVES

After completing this chapter, you should be able to:

1. Determine the polarity of voltage across circuit components.
2. Trace the flow of electrons through series, parallel, and series parallel (combination) circuits.
3. Determine distribution of voltage drops and equivalent values of resistance in all parts of a combination circuit.
4. Mathematically calculate voltage drop.
5. Determine the proper wire sizes needed to reduce voltage drop due to line loss to meet *National Electrical Code*® requirements.

■ GLOSSARY

Ampacity The maximum current the cable can carry safely without exceeding its temperature rating.

Eddy currents Usually undesirable currents that are set up in metal structures, such as conduits or motor cores. Eddy currents cause losses.

Fault A failure of some type in the electrical system, usually a short circuit.

Polarity The relationship (minus to plus or plus to minus) between opposite ends of a circuit load or resistance.

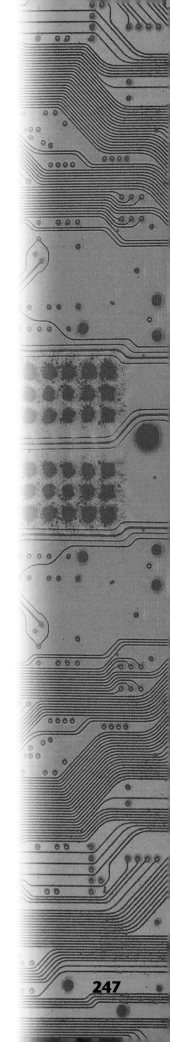

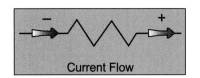

FIGURE 21–1 Polarity according to direction of current flow.

■ DETERMINING VOLTAGE POLARITY

When current flows through a load or resistor, it takes on a **polarity**. Since electrons flow from negative to positive, it is obvious that when electrons flow through a resistor, a voltage drop ($I \times R$) is developed across the resistor, with one end becoming negative and the other positive. A diagram of a load, with polarity markings indicated, shows which way electrons will flow. Analyzing electrical circuits correctly often requires the use of laws and theorems that depend on the principle of polarity. You must know how to determine the polarity across each component in order to solve these types of problems. In Figure 21–1, for example, the current through the resistor is flowing from left to right. The left side of the resistor, where the current enters, is more negative than the right side.

In a circuit, the polarity of components is determined by tracing the current path, starting at the negative side of the power source. When a component is encountered, a negative sign is placed on the side the current enters and a positive sign on the side the current exits. Continue doing this until all the components are marked with polarity signs. A completed circuit is shown in Figure 21–2.

Notice that where the current exits R_1 is positive and where it enters R_2 is negative. This does not mean that the negative side of R_2 is more negative than the positive side of R_1—they are actually the same potential. The polarity signs show how ends of the same component compare to each other. If a voltmeter were used to measure the difference between points A and B, it would read zero since there is no difference in potential.

In the same respect, a bird sitting on a high-voltage transmission line is not shocked because there is effectively no difference in potential between its feet. Since every wire has resistance, there would be minute voltage, depending on the wire size, and a minute circuit current. The bird is sitting between two points that have very little potential difference (note that the + side of R_1 and the − side of R_2 in Figure 21–2 are the same potential).

Later, when you learn about AC circuits, you will find that polarity and polarity markings are used there as well. In the AC case, however, the polarity markings indicate locations that are substantially in phase with one another.

FIGURE 21–2 Polarity of components in a circuit.

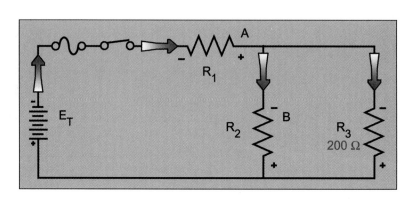

■ CHOOSING THE RIGHT CONDUCTOR

Many factors must be taken into account to prevent cables from being underrated when wiring buildings or other facilities that use electrical cabling. Underrating a cable can cause the insulation for the cable to deteriorate, even to the point of it becoming a fire hazard. The *National Electric Code*® (*NEC*®) specifies the ratings for wires and cabling based on application of the physical law governing electrical circuits. The code must always be followed when installing the cables to ensure that electrical fires do not start because of overheating conductors or devices. Also, the wires must be chosen so that rated equipment can operate at the correct voltage and current. Systems with too much inherent resistance will reduce both the current to the load and the efficiency.

21.1 Ampacity

The **ampacity** of a cable or wire is the maximum current the cable can carry safely without exceeding its temperature rating. The *NEC*® requires that branch circuit conductors have an ampacity greater than or equal to the noncontinuous load plus 125% of the continuous load.

21.2 Conductor Voltage Drop

In your studies, you have used circuits for calculation of resistance, voltage, and current assuming that the cables and wire had zero resistance and therefore zero voltage drop. In the real world, all cables and wires have real resistance and create a voltage drop. To ensure that the efficiency of the electrical system is as high as possible, the *NEC*® requires that cables and wires be chosen so that the voltage drop from either feeders or branch circuits to the farthest outlet does not exceed 3%. Also, it recommends that the combined voltage drop for branch circuits and feeders not exceed 5%. These two ratings ensure a reasonable efficiency for the total circuit. Figure 21–3 shows an example of feeder and branch lines.

EXAMPLE 1

a. A branch circuit has an applied voltage of 240 volts. What would be the allowable voltage drop?
b. What would be the consumed power by a 240-V/750-W heater on the branch?

Solution:

$$E_{VD} = E_{supply} \times 3\% = 240 \times .03 = 7.2 \text{ V}$$

The actual voltage applied is

$$E_{app} = E_{supply} - E_{VD} = 240 - 7.2 = 232.8 \text{ V}$$

The resistance of the heater is

$$R = \frac{E^2}{P} = \frac{240^2}{750} = 76.8 \ \Omega.$$

FIGURE 21–3   Feeder and branch circuit voltage drops.

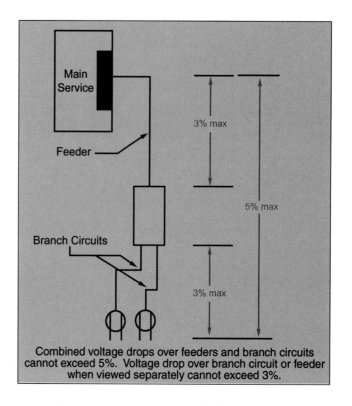

Combined voltage drops over feeders and branch circuits cannot exceed 5%. Voltage drop over branch circuit or feeder when viewed separately cannot exceed 3%.

So, the power consumed by the heater is

$$P = \frac{E_{VD}^2}{R} = \frac{232.8^2}{76.8} = 705.68 \text{ W}$$

21.3 Determining Conductor Resistance

The code requirements in the *NEC*® assume that the length of a wire will not substantially affect the ampacities of the wires in a system. In systems with very long wires, this may not be true. Therefore, it may be necessary to compute the size of the wire in systems with very long wires.

The resistance of a wire is affected by four factors:

1. The type of material (some materials are more resistive than others)
2. The diameter of the wire
3. The length of the wire (making a wire longer is like adding resistors in series)
4. The temperature of the wire

When considering the length of a wire and resistance, the standard for resistance in the English system is the mil foot. A mil foot is a length of wire 1 mil in diameter and 1 foot long. A table showing the types of materials is shown as Table 21–1. The table shows the resistance per mil for the material in the first column and the amount of resistance change per degree change in temperature.

Table 21.1   Resistivity (K, ρ) of Materials

Material	K (ρ) at 68°F (20°C) American (English)	Metric	Resistance Temperature Coefficient (Ω/°C)
Aluminum	17.7	.0265	.004308
Copper	10.4	.0168	.004041
Lead	126	.22	.0043
Mercury	590	0.98	.00088
Nichrome	600.0	1	.00017
Platinum	66	.106	.003729
Silver	9.7	.0159	.003819
Tungsten	33.8	.056	.004403

The voltage drop E_{vd} across any resistance (load or wire) is equal to $I \times R$, and the resistance of wire is found by using the formula

$$R = \frac{K \times L}{A}$$

where R is the resistance of the wire, K is the ohms per mil foot (or ohms per millimeter-meter in SI units), L is the length of wire in feet (or meters in SI units), and A is the area in circular-mils (CM) (or square meters in SI). This formula shows that resistance is fixed and does not exist as a condition or result of voltage or current. Different variations of the equation can be used.

Table 21–2 is a slightly modified version of Table 3–3. Both tables provide size and resistance information; however, Table 21–2 shows values for a variety of types of wire (stranded and solid) as well as different materials (coated copper, uncoated copper, and aluminum).

EXAMPLE 2

What is the resistance of a piece of 8-gauge copper wire 1,000 feet long? The temperature of the wire is 20°C (68°F).

Solution:

$$R = \frac{K \times L}{A}$$

Substitute in the formula from Table 21–1, $K = 10.4\ \Omega$ per foot at 20°C, $L = 1,000$ feet and, from Table 21–2, A (CM) for a #8-gauge wire is 16,510 CM:

$$R = \frac{(10.4)(1,000)}{16,510} = 0.63\ \Omega$$

Table 21–2   American Wire Gauge

| Size AWG/ kcmil | Area Cir. Mills | Conductors | | | | DC Resistance at 75°C (167°F) | | |
| | | Stranding | | Overall | | Copper | | Aluminum |
		Quantity	Diam. In.	Diam. In.	Area In.^{2}	Uncoated ohm/kFT	Coated ohm/kFT	ohm/kFT
18	1620	1	—	0.040	0.001	7.77	8.08	12.8
18	1620	7	0.015	0.046	0.002	7.95	8.45	13.1
16	2580	1	—	0.051	0.002	4.89	5.08	8.05
16	2580	7	0.019	0.058	0.003	4.99	5.29	8.21
14	4110	1	—	0.064	0.003	3.07	3.19	5.06
14	4110	7	0.024	0.073	0.004	3.14	3.26	5.17
12	6530	1	—	0.081	0.005	1.93	2.01	3.18
12	6530	7	0.030	0.092	0.006	1.98	2.05	3.25
10	10380	1	—	0.102	0.008	1.21	1.26	2.00
10	10380	7	0.038	0.116	0.011	1.24	1.29	2.04
8	16510	1	—	0.128	0.013	0.764	0.786	1.26
8	16510	7	0.049	0.146	0.017	0.778	0.809	1.28
6	26240	7	0.061	0.184	0.027	0.491	0.510	0.808
4	41740	7	0.077	0.232	0.042	0.308	0.321	0.508
3	52620	7	0.087	0.260	0.053	0.245	0.254	0.403
2	66360	7	0.097	0.292	0.067	0.194	0.201	0.319
1	83690	19	0.066	0.332	0.087	0.154	0.160	0.253
1/0	105600	19	0.074	0.372	0.109	0.122	0.127	0.201
2/0	133100	19	0.084	0.418	0.137	0.0967	0.101	0.159
3/0	167800	19	0.094	0.470	0.173	0.0766	0.0797	0.126
4/0	211600	19	0.106	0.528	0.219	0.0608	0.0626	0.100
250	—	37	0.082	0.575	0.260	0.0515	0.0535	0.0847
300	—	37	0.090	0.630	0.312	0.0429	0.0446	0.0707
350	—	37	0.097	0.681	0.364	0.0367	0.0382	0.0605
400	—	37	0.104	0.728	0.416	0.0321	0.0331	0.0529
500	—	37	0.116	0.813	0.519	0.0258	0.0265	0.0424
600	—	61	0.099	0.893	0.626	0.0214	0.0223	0.0353
700	—	61	0.107	0.964	0.730	0.0184	0.0189	0.0303
750	—	61	0.111	0.998	0.782	0.0171	0.0176	0.0282
800	—	61	0.114	1.030	0.834	0.0161	0.0166	0.0265
900	—	61	0.122	1.094	0.940	0.0143	0.0147	0.0235
1000	—	61	0.128	1.152	1.042	0.0129	0.0132	0.0212
1250	—	91	0.117	1.289	1.305	0.0103	0.0106	0.0169
1500	—	91	0.128	1.412	1.566	0.00858	0.00883	0.0141
1750	—	127	0.117	1.526	1.829	0.00735	0.00756	0.0121
2000	—	127	0.126	1.632	2.092	0.00643	0.00662	0.0106

This resistance formula can also be used similar to the Ohm's law formula. Using Ohm's law, you can calculate R, E, P, or I. In the same way, you can use this formula to determine the following:

$$R = \frac{K \times L}{A} \quad \text{for resistance of a wire}$$

$$K = \frac{R \times A}{L} \quad \text{for the resistivity of a wire}$$

$$L = \frac{R \times A}{K} \quad \text{for the length of a wire}$$

$$A = \frac{K \times L}{R} \quad \text{for the size of a wire}$$

21.4 Calculating Voltage Drop

Another aspect of constructing electrical systems is taking into account the voltage drop of the system. By substituting the resistance formula into Ohm's law, we get $E_{VD} = \frac{(I \times K \times L)}{A}$. When calculating voltage drop in a circuit (or feeder), 2L is often used because a circuit usually requires two conductors. The formula then becomes

$$E_{VD} = \frac{(I \times K \times 2L)}{A}$$

A simple transposition of the formula makes it possible to solve for wire sizes if the current and permitted voltage drop are known:

$$A = \frac{(I \times K \times 2L)}{E_{VD}}$$

EXAMPLE 3

A certain load is being supplied from a feeder 300 feet away. The feeder uses #6-gauge copper wire, and the load requires 30 A (see Figure 21–4). What is the voltage drop across the conductors when the load is operating? What is the percentage of voltage drop for the load? Is this greater than the *NEC*®-suggested voltage drop for circuit lines if the load is supplied by a 240-V feeder?

$$E_{VD} = \frac{(I \times K \times 2L)}{A} = \frac{30 \times 10.4 \times 2 \times 300}{26,240} = 7.13 \text{ V}$$

$$\% \text{ VD} = \frac{E_{VD}}{E_{supply}} \times 100 = \frac{7.13}{240} \times 100 = 2.97\%$$

No, this does not exceed the requirement of no more than 3% voltage drop.

FIGURE 21–4   Voltage drop and wire resistance calculation for #6 copper feeder.

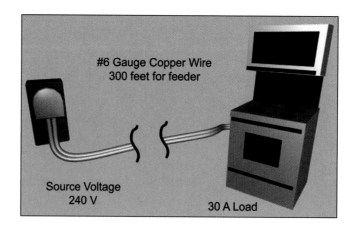

21.5 Special Cases

Parallel Conductors

There are times when using a single set of conductors to provide power to a load can become impractical:

1. When a set of conductors with sufficient ampacity to supply a load would be too large or difficult to handle
2. When the size of the required conductors is so large that only one conductor would fit in a raceway or conduit (which is prohibited by the *NEC*® because of the effects of induction generating heat within the conduit material)

When these situations occur, one solution is to use multiple sets of smaller conductors, connected in parallel, in place of the one set of larger conductors. The basic requirements when using parallel conductors are that the individual conductor size cannot be smaller than 1/0 and that the combined area of the conductors being used in parallel must be equal to or greater than the area of the original larger conductor.

EXAMPLE 4

A large motor, located approximately 2,500 feet from the source, requires a 750-kcmil conductor feed. A decision is made to run four parallel sets of 4/0 conductors in place of the larger 750-kcmil feeder.

Is this an acceptable solution?

Solution:

Based on Table 21–2,

750-kcmil conductor = 750,000 cmil

4/0 conductor = 211,600 cmil

four 4/0 conductors in parallel = 846,400 cmil

Since the four parallel conductors have a combined area of 846,600 CM and the minimum requirement was 750,000, the parallel sets of 4/0 would be an acceptable replacement to the 750-kcmil set of conductors.

The *NEC*® requires five conditions to use parallel conductors in place of a single conductor:

1. Each conductor must have the same area.
2. Each conductor must have the same length.
3. Each conductor must be of the same material (copper to copper or aluminum to aluminum).
4. Each conductor must have the same insulation.
5. Each conductor must be connected/terminated in the same way.

Eddy Currents

When three-phase AC loads are large and parallel conductors must be used to carry the large current, it is important that conductors of each of the three phases be installed in each conduit. The reason for this is because of the alternating, or "back and forth," motion of the AC, which produces expanding and collapsing magnetic fields. These fields are induced into the metal making up the walls of the conduit in which the conductors are installed. Since the conduit is similar to the core of an electromagnet, a current is induced into the conduit. Heat builds up because of the strength and energy of **eddy currents** (currents induced in metal components surrounding or near conductors). To prevent this, equal numbers of conductors for each phase must be installed in each conduit. The magnetic fields of different phases will cancel out the magnetic field and eliminate the eddy current effects (see Figure 21–5).

High-Voltage Testing

When conduits or raceways are used for wire installation, you need to check for any of the conductors being shorted to the ground or any of

FIGURE 21–5 Parallel conductors in a conduit.

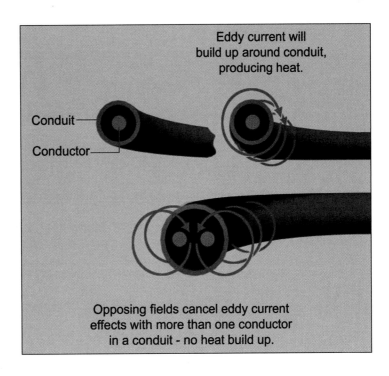

the conductors that may be shorted to each other before applying power to the circuit. Often these types of **faults** are high-resistance faults. This means that you need to be able to measure megohms of resistance (remember, 1 megohm = 1,000,000 ohms) using an ohmmeter. Not only do you need an ohmmeter that can measure millions of ohms, but you also need to be able to supply enough voltage to check the insulation against breakdown. Regular ohmmeters do not have the capacity to provide this kind of voltage.

When performing high-voltage tests, a special meter called a megohmmeter, or Megger®, is used. The Megger® is a special ohmmeter that measures high resistance using voltages between 250 and 5,000 volts, depending on the voltage rating of the cable being tested. When the Megger® is placed across two conductors in a circuit and the meter applies the correct voltage, an accurate reading of the resistance in megohms between the two conductors can be read. Generally, these readings are recorded so that they can be compared with previous readings. If the insulation resistance is too low, the conductor must be replaced. Follow the cable manufacturer's recommendations for acceptable values.

To check for grounds using the Megger®, one meter lead is connected to the conduit or raceway and the other to a load conductor. The conductor should be tested at the rated load voltage or slightly higher. Each load conductor in the conduit or raceway needs to be tested in this way.

■ SUMMARY

Polarity is an important concept for the student of electricity to understand. Polarity must be observed when connecting many electrical and electronic devices to power sources. Failure to do so can result in damage or destruction to the devices used. The ability to effectively use several laws and theorems concerning electrical principles depends on the proper identification of polarity.

Polarity is defined with the direction of current (electron) flow. Using the electron theory, current is the flow of electrons. Current flows from negative (a point with an excess of electrons) to positive (a point with a deficiency of electrons) in the external circuit. If this is true, then as electrons pass through a resistance or load, the end from which they enter must be negative; and the end through which they leave must be positive.

How is the polarity of the voltage drop across a load determined? The first step is to determine the direction of current flow. This is easy if you remember that current flows from the negative side of the voltage source, through the load, to the positive side of that source. Once the direction of current has been established, the second step is to indicate the polarity of the voltage across the load.

The formula $R = \dfrac{K \times L}{A \, (\text{cmil})}$, shows that resistance in a conductor is fixed and does not exist as a condition or result of voltage or current. The different variations of the equation that can be used are the following:

$R = \dfrac{K \times L}{A}$ for resistance of a wire

$K = \dfrac{R \times A}{L}$ for the resistivity of a wire

$L = \dfrac{R \times A}{K}$ for the length of a wire

$A = \dfrac{K \times L}{R}$ for the size of a wire

Another variation of the formula is used to calculate the voltage drop in a circuit:

$$E_{VD} = \frac{(I \times K \times 2L)}{A}$$

There are two times when parallel conductors are needed: when the size of a single conductor is too large to be handled easily and when multiple conductors for each phase of three-phase loads need to be in the same conduit. Other special requirements for testing high-voltage conductors in conduits can be accomplished using a Megger®. This special meter tests for shorts and grounds and for conductor insulation at high voltages.

■ REVIEW QUESTIONS

1. Which way does electron current flow?
2. What is polarity?
3. Discuss what happens if you hook up a meter with reverse polarity.
4. What is the maximum voltage drop allowed by code on a feeder?
5. What is the maximum VD allowed by the *NEC*® on feeders and branch circuit combined?
6. Discuss the unit of cross-sectional area for wires in both American and SI units.
7. What is a circular mil?
8. What is the formula for calculating voltage drop?
9. What is a megohmmeter?

■ PRACTICE PROBLEMS

1. What is the voltage drop for a circuit that feeds a 40-amp load, is 600 feet away from the source, and is fed with a #6 copper at 208 volts? Based on your calculations, is the voltage drop acceptable?
2. What size conductor would it take to meet the requirements of the *NEC*® based on the calculated values from Example 1?

Applying the Principle of Superposition

■ OUTLINE

■ OVERVIEW

The superposition theorem is one of the many useful tools that can be used in circuit analysis. It is so useful because it applies Ohm's law to circuits that have more than one voltage source. In simple terms, we can calculate the effect of one voltage source at a time and then superimpose the results of all the sources. You can also use the superposition theorem to calculate the current flow through any circuit branch containing more than one voltage source.

■ OBJECTIVES

After completing this chapter, you should be able to:

1. State the steps necessary to apply the principle of superposition.
2. Apply the theory of superposition to solve for multiple voltage source circuits.

■ GLOSSARY

Current source   A source that keeps its output voltage constant regardless of the load applied. In the early days of electric power, street lighting circuits were often series circuits supplied with constant current transformers. In modern power systems, current sources are rarely, if ever, encountered. Current sources are more commonly used in electronics systems.

Voltage source   A source that keeps its output voltage constant regardless of the load applied. Such a device does not exist in reality; however, batteries and other such power sources approximate them.

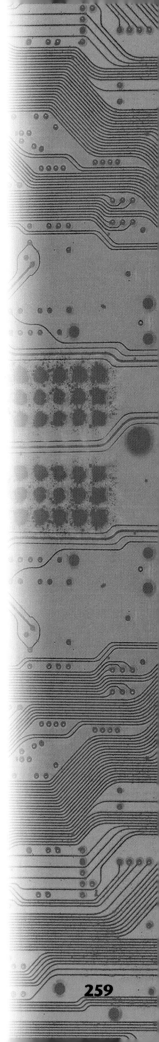

■ THE SUPERPOSITION THEOREM

22.1 Definition

The voltage or current in any element resulting from several sources acting together is the sum of the voltages or currents resulting from each source acting alone. In order to produce the effects for a single source at a time, all other sources must be turned down to zero while the first source effects are calculated. This means disabling the other sources so that they cannot generate voltage or create current. This has to be done without changing the resistance of the circuit. You can do this by assuming that the voltage sources not being measured are shorted across their terminals.

22.2 Superposition Theorem Steps

Example 1 takes a step-by-step approach for using the superposition theorem when solving for the effect of multiple voltage sources in a circuit (see Figure 22–1).

EXAMPLE 1

Find the voltage drop and current flow of R_3 in Figure 22–1.

Solution:

Step 1

Reduce all but one **voltage source** to zero. In this example (as seen in Figure 22–1), you will remove the V_2 voltage source and replace it with a jumper as seen in Figure 22–2. By inserting the jumper in place of the V_2 source, you have removed the effects of the V_2 source on the circuit; essentially, you have turned the source to zero by putting a short circuit across it.

If this were a **current source**, you would reduce the current to zero by creating an open circuit at the current source. To create the open circuit at the current source, the current source would be removed from the circuit, but this time you would not install the jumper in place of the current source.

Next, calculate the total resistance of the new circuit as shown in Figure 22–2 using Ohm's law.

FIGURE 22–1   A circuit with two sources.

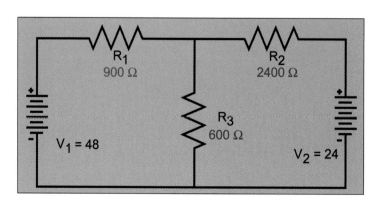

FIGURE 22–2   Step 1: The first source is turned to zero.

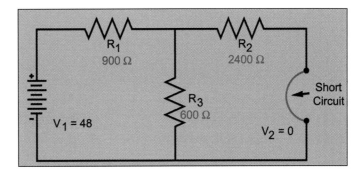

First, reduce resistors R_2 and R_3 (which are now in parallel) to an equivalent resistance that is called $R_{2,3}$:

$$R_{2,3} = \frac{R_2 \times R_3}{R_2 + R_3} = \frac{2{,}400 \times 600}{2{,}400 + 600} = 480 \ \Omega$$

Now calculate the equivalent resistance for resistors R_1 and $R_{2,3}$. Since they are in series, you know that

$$R_T = R_1 + R_{2,3} = 900 + 480 = 1{,}380 \ \Omega$$

Now that the total resistance is known, calculate the total current for the circuit:

$$I_T = \frac{E_T}{R_T} = \frac{48}{1{,}380} = .0348 \ \text{A}$$

Since the total current and resistance are known, the voltage drop across the parallel resistors R_2 and R_3 can be found:

$$E_{2,3} = I_T \times R_{2,3} = 0.384 \times 480 = 16.7 \ \text{V}$$

Finally, calculate the current through R_3:

$$I_{R_3} = \frac{E_{R_{2,3}}}{480} = \frac{16.7}{480} = .0278 \ \text{A}$$

Look at Figure 22–3. Note the polarity of the voltage drop and the direction of the current flow for R_3.

FIGURE 22–3   Step 1: Results of the first step with the first source turned to zero.

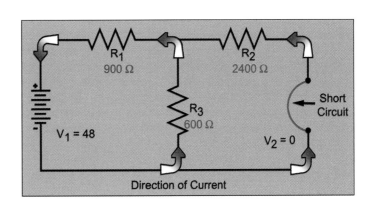

Direction of Current

Step 2

The next step is to repeat the previous steps using V_2 as the only voltage source. To do this, remove the jumper and reinstall V_2 back into the circuit. Then remove V_1 and replace it with a jumper, just as you did in step 1. Figure 22–4 shows the newly configured circuit.

Now calculate the total resistance in the redrawn circuit:

$$R_T = R_2 + \left(\frac{1}{\frac{1}{R_1} + \frac{1}{R_3}}\right) = 2,400 + \left(\frac{1}{\frac{1}{900} + \frac{1}{600}}\right) = 2,760 \ \Omega$$

Now that the total resistance is known, calculate the total current for the circuit:

$$I_T = \frac{E_T}{R_T} = \frac{24}{2,760} = .0087 \ A$$

Since the total current and resistance are known, the voltage drop across the parallel resistors R_1 and R_3 can be found:

$$E_{1,3} = I_T \times R_{1,3} = .0087 \times 360 = 3.13 \ V$$

Finally, the current through R_3 can now be calculated:

$$I_{R_3} = \frac{E_{1,3}}{R_3} = \frac{3.13}{600} = .0052 \ A$$

Look at Figure 22–5. Note that the polarity of the voltage drop and the direction of the current flow for R_3 are the same as in Figure 22–3 after step 1.

FIGURE 22–4   Step 2: The first source is restored, and the second source is turned to zero.

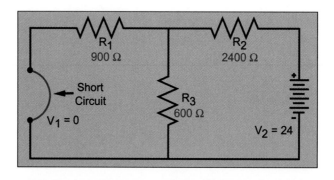

FIGURE 22–5   Step 2: Results of the second step with second source turned to zero.

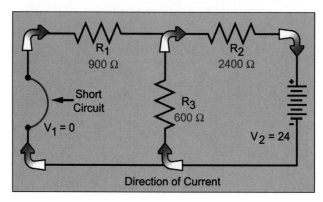

Step 3

The last step is to find the algebraic sum of the two currents and voltage drops in R_3. Look at Figures 22–3 and 22–5. Since both currents are flowing in the same directions and both voltage drops have the same polarities, the voltages and currents are simply added together:

$$I_{R_3} = I_{R_3\text{-step1}} + I_{R_3\text{-step2}} = .0052 + .0278 = .0330 \text{ A}$$

$$E_{R_3} = E_{R_3\text{-step1}} + E_{R_3\text{-step2}} = 3.13 + 16.7 = 19.83 \text{ V}$$

EXAMPLE 2

Using Figure 22–6, solve for the total current flow through R_3. Notice that this circuit has both a voltage source and a current source.

Solution:

Step 1

Reduce the current source to zero by opening the circuit at the terminals of the current source. Figure 22–7 shows the redrawn circuit.
   Now calculate the total resistance in the redrawn circuit:

$$R_\text{T} = R_1 + R_3 = 900 + 600 = 1,500 \text{ }\Omega$$

Note that R_2 did not add to the total resistance seen by V_1 because no current could flow through the open part of the circuit.

FIGURE 22–6   Superposition theorem—current source.

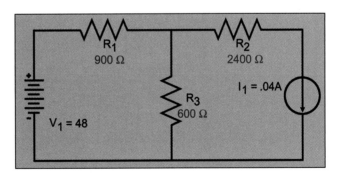

FIGURE 22–7   Redrawn with current source removed (set to zero).

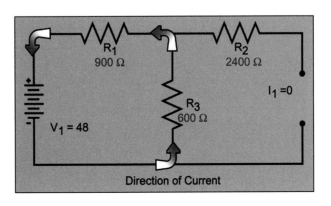

Now that the total resistance is known, calculate the total current for the circuit:

$$I_T = \frac{E_T}{R_T} = \frac{48}{1,500} = .032 \text{ A}$$

Since the total current and resistance is known, the voltage drop across the resistors R_1 and R_3 can be found:

$$E_1 = I_T \times R_1 = .032 \times 900 = 28.8 \text{ V}$$
$$E_3 = I_T \times R_3 = .032 \times 600 = 19.2 \text{ V}$$

Step 2

Now reverse the process for I_1. Short out V_1 and reduce its voltage to zero. Figure 22–8 shows how this will affect the circuit.

Reduce the voltage source to zero. Replacing the voltage source with a short circuit does this (see Figure 22–8).

Now calculate the total resistance in the redrawn circuit. First, solve for the parallel branch made up of resistors R_1 and R_3 using Ohm's law:

$$R_{1,3} = \frac{R_1 \times R_3}{R_1 + R_3} = \frac{900 \times 600}{900 + 600} = 360 \text{ }\Omega$$

Since the new $R_{1,3}$ is in series with R_2, you can easily calculate the total resistance R_T value:

$$R_T = R_{1,3} + R_2 = 360 + 2,400 = 2,760 \text{ }\Omega$$

Now that the total resistance is known, calculate the total voltage for the circuit:

$$E_T = I_1 \times R_T = .04 \times 2,760 = 110.4 \text{ V}$$

Since the total current and resistance are known, the voltage drop across the parallel resistors R_1 and R_3 can be found:

$$E_{1,3} = I_T \times R_{1,3} = .04 \times 360 = 14.4 \text{ V}$$

FIGURE 22–8 Redrawn circuit with voltage source removed.

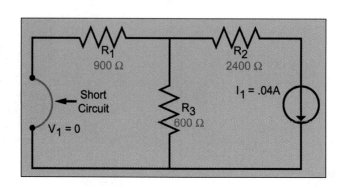

Finally, the current through R_1 and R_3 can now be calculated:

$$I_1 = \frac{E_{1,3}}{R_1} = \frac{14.4}{900} = .016 \text{ A}$$

$$I_3 = \frac{E_{1,3}}{R_3} = \frac{14.4}{600} = .024 \text{ A}$$

Look at Figure 22–9. Note the polarity of the voltage drop and the direction of the current flow for R_1 and R_3.

Step 3

The last step is to find the algebraic sum of the two currents and voltage drops in R_1 and R_3. Look at Figures 22–7 and 22–9. Since the two currents that flow through R_1 are in opposite directions and the voltage drops have the opposite polarities, the voltages and currents are algebraically added together (subtracted), and the larger of the currents and voltage drops determine the direction of polarity and current flow for R_1:

$$I_{1 \text{ total}} = I_{T \text{ (with current source open)}} - I_{1 \text{ (with } V_1 \text{ shorted)}}$$
$$I_{1 \text{ total}} = .032 - .016 = .016 \text{ A}$$
$$V_{1 \text{ total}} = V_{1 \text{ (with current source open)}} - V_{\text{comb (with } V_1 \text{ shorted)}} = V_{1 \text{ total}} =$$
$$28.8 - 14.4 = 14.4 \text{ V}$$

Note that the current flow and voltage drops for I_3 are in the same direction and of the same polarity. This allows us to add the currents and voltage drops for the total effect on R_3:

$$I_{3 \text{ total}} = I_{3 \text{ (with current source open)}} + I_{3 \text{ (with } V_1 \text{ shorted)}} = .032 + .024 = .056 \text{ A}$$
$$V_{3 \text{ total}} = V_{3 \text{ (with current source open)}} + V_{1,3 \text{ (with } V_1 \text{ shorted)}} = 19.2 + 14.4 = 33.6 \text{ V}$$
$$V_{3 \text{ total}} = I_T \times R_3 = .056 \times 600 = 33.6 \text{ V}$$

FIGURE 22–9   Circuit current flow for Example 2.

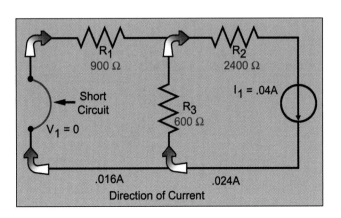

22.3 Superposition Theorem Requirements

There are two basic circuit requirements necessary for the use of the superposition theorem:

1. All the components must be linear and bilateral. Linear means that the current is proportional to the applied voltage. Bilateral means that the calculated amount of the current through a component or circuit would be the same even if the voltage source applied were connected in reverse polarity.

2. The circuit components are usually "passive." Passive components do not amplify or rectify. Examples of linear and bilateral components include capacitors, resistors, and air-core inductors. Examples of active components include transistors, semiconductor diodes, and tubes.

■ SUMMARY

The superposition theorem is very useful in solving complex circuits with more than one voltage or current source. To use the theorem, the circuit has to have passive components that are linear and bilateral. Normally, resistors, capacitors, and air-core inductors meet these requirements.

The steps in applying the theorem are the following:

1. Reduce all but one voltage and/or current source to zero. Calculate the total resistance, voltage, and current for this "redrawn" circuit.

2. Repeat step 1 for each of the other sources in the circuit.

3. Combine the resulting voltages and currents through the different resistances. This is done algebraically. The results are the combined voltage and current effects on the various resistances within the circuits.

■ REVIEW QUESTIONS

1. What is the simple definition of the superposition theorem?

2. What are the steps that must be taken when using the superposition theorem to solve a problem with two voltage sources?

3. What are the steps that must be taken when using the superposition theorem when you have a voltage source and current source?

■ **PRACTICE PROBLEM**

1. Use the superposition theorem to calculate the voltage drops and current flows for all the resistors in the following circuit.

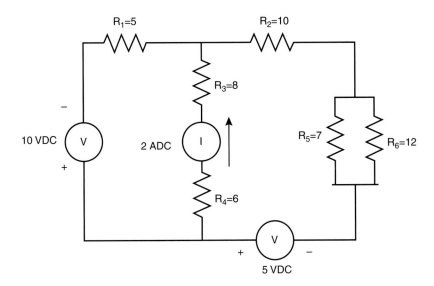

chapter **23**

Using Kirchhoff's Laws to Solve DC Circuits

■ **OUTLINE**

■ OVERVIEW

Kirchhoff's laws provide you with the tools that are necessary to solve more complex circuits than is possible using Ohm's law. Applying these laws allows you to gain a better working knowledge of circuit operations necessary for solving multiple-source circuits. Kirchhoff's laws are *not* replacing Ohm's law but are going beyond the basics and giving you new skills in circuit analysis.

Kirchhoff's two laws are the following:

1. The algebraic sum (Σ) of the currents entering and leaving any node (junction point) is zero.
2. The algebraic sum (Σ) of the voltages around any closed path is zero.

This chapter will introduce you to the two laws and provide direction and practice in applying them to the solving of complex, combination circuits. In many cases, the use of Kirchhoff's laws is more direct than the superposition theorem.

■ OBJECTIVES

After completing this chapter, you should be able to:

1. State Kirchhoff's laws both verbally and mathematically.
2. Apply Kirchhoff's laws to solve for circuit variables in complex circuits.

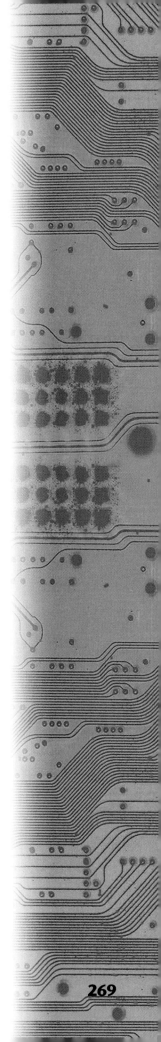

■ THE TWO LAWS

23.1 Kirchhoff's Current Law

Kirchhoff's current law states that the sum of any currents entering and leaving any given point in a circuit must be equal to zero. When summing the values of current entering and leaving a point within the circuit, the currents entering the point are positive, while the currents leaving the point are negative. Looking at Figure 23–1, Kirchhoff's current law can be written algebraically as

$$I_{R_1} + I_{R_2} + I_{R_3} = 0$$

As you look at Figure 23–1, you might realize that the current flow through R_3 is actually flowing away from node E, not into it as shown. This is the beauty of both of Kirchhoff's laws: As long as you are consistent with your assumptions, the direction of current flow will work out correctly when you do the algebraic solution. This means that, for example, if you always assume your currents to be into each node, their actual direction will be sorted out by the algebra. For example, in Figure 23–1 you have assumed that the current through R_3 is flowing into node E. When this circuit is solved, I_{R_3} will have a negative sign showing that you assumed the wrong direction.

Of the two laws, you will probably use Kirchhoff's voltage law more often than Kirchhoff's current law. The same principle applies in the voltage law, as you will see in the next section.

23.2 Kirchhoff's Voltage Law

Refer to Figure 23–1 again. The circuit has three possible loops (closed paths): ABCFED, ABED, and CBEF. In Kirchhoff's voltage law, you need to select loops so that every component is included in at least one; consequently, you need only the last two, ABED and CBEF, to analyze the entire circuit. Remember that you need to assume only as many loops as you need to include every component. In some cases, one or more of the components will be included in more than one of the loops. Figure 23–2 shows that resistor R_3 is included in both the i_1 loop and the i_2 loop.

FIGURE 23–1   Kirchhoff's current law.

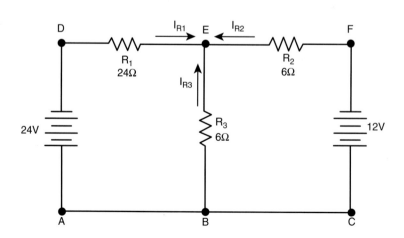

FIGURE 23-2   Kirchhoff's voltage law example showing the i_1 and i_2 loops.

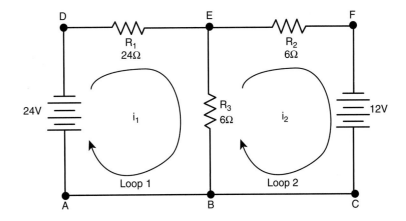

FIGURE 23-3   The voltage drops selected for Figure 23-3.

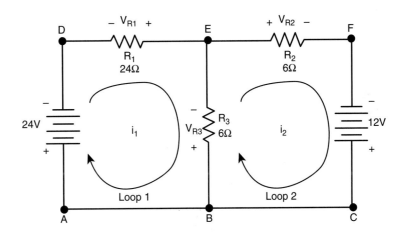

Notice that as each current flows around its own loop, it will create a voltage drop across the resistors. In resistor R_3, the voltage drop will be caused by both i_1 and i_2. Although the choice of polarities for the voltage drops is not critical (as stated earlier), it makes sense to select them logically. Consider Figure 23-3.

In Figure 23-3, each voltage drop is labeled and assigned a polarity. The polarities are chosen as follows:

1. V_{R_1} is chosen with its negative end located where the i_1 current flows in. This is a reasonable guess and is the way that most of the polarities should be chosen.

2. V_{R_2} is chosen with its positive end located where the i_2 current flows in. This is opposite the way that you would normally choose it. The polarity is chosen for this example only and will help illustrate that the algebra will eventually sort out the correct polarity.

3. V_{R_3} is shared between two loops; therefore, its polarity cannot be assumed. In such cases, the polarity is often picked on the basis of the first loop you encounter as you analyze—in this case, it assumes the polarity of loop 1 (i_1).

Since you have assumed clockwise rotation for i_1 and i_2, the Kirchhoff voltage equations can be developed by starting at the source and adding voltages in a clockwise direction for each loop. Be careful to observe the signs that you have assumed:

For loop 1,

$$+24 - V_{R_1} - V_{R_3} = 0 \tag{1}$$

For loop 2,

$$-12 + V_{R_3} + V_{R_2} = 0 \tag{2}$$

Notice that for each loop, you start with the source. This is not totally necessary, but it is easy to remember. If there is more than one source, you can start with either one. As you go around the loop in the same direction as the current, you add or subtract each voltage, depending on the sign that you encounter. In Equation 1, you are moving clockwise, and you encounter the positive sign for the 24-V battery first; therefore, it is listed as positive. The other voltages are identified in the same way.

Equations 1 and 2 can be simplified even more by moving the sources to the right side of the equation and multiplying Equation 1 by -1. This gives

$$V_{R_1} + V_{R_3} = 24 \tag{3}$$

and

$$V_{R_3} + V_{R_2} = 12 \tag{4}$$

■ APPLYING KIRCHHOFF'S VOLTAGE LAW

EXAMPLE 1

Equations 3 and 4 define the voltages; however, even these equations are not yet enough. Notice that you have three unknowns (V_{R_1}, V_{R_2}, and V_{R_3}) and only two equations. Fortunately, you can simplify by substituting the Ohm's law equivalents for the three unknowns. From Ohm's Law, you know that

$$V_{R_1} = i_1 \times R_1$$

$$V_{R_2} = -i_2 \times R_2$$

$$V_{R_3} = i_1 R_3 - i_2 R_3$$

Notice that the value for V_{R_2} is negative. This is because the current i_2 is flowing into the positive end of R_2 as we have defined it. This means that the actual value will be minus the current times the resistance as shown earlier. Notice also that since i_1 flows into the negative end of R_3 and i_2 flows into the positive end, the total voltage will be as shown. Substituting these values into Equations 3 and 4 gives

$$i_1 R_1 + (i_1 - i_2)R_3 = 24 \tag{5}$$

and

$$(i_1 - i_2)R_3 - i_2 R_2 = 12 \tag{6}$$

Simplifying and collecting terms gives

$$i_1(R_1 + R_3) - i_2(R_3) = 24 \tag{7}$$

and

$$i_1 R_3 - i_2(R_2 + R_3) = 12 \tag{8}$$

Substituting the known values for the three resistors gives

$$30i_1 - 6i_2 = 24 \tag{9}$$

and

$$6i_1 - 12i_2 = 12 \tag{10}$$

Notice that both Equation 9 and Equation 10 can be divided by 6, which leaves

$$5i_1 - i_2 = 4 \tag{11}$$

and

$$i_1 - 2i_2 = 2 \tag{12}$$

You now have two equations with two unknowns to solve for i_1 and i_2. There are several ways to solve for these two unknowns. The easiest would be to use a scientific calculator, computer spreadsheet, or other software that has the ability to solve them. In fact, when working on very complicated circuits with three or more loops, this may be the best practical approach.

Solution:
In this simple example, you can use the fairly simple approach of eliminating one of the unknowns and solving for the other. First, multiply Equation 11 by 2 to produce Equation 13:

$$2 \times (5i_1 - i_2 = 4) = 10i_1 - 2i_2 = 8 \tag{13}$$

You can then subtract Equation 13 from Equation 12 to get

$$(i_1 - 2i_2 = 2) - (10i_1 - 2i_2 = 8) \tag{14}$$

Expanding and collecting terms gives

$$i_1 - 10i_1 - 2i_2 + 2i_2 = -6$$

which further simplifies to

$$-9i_1 = -6 \Rightarrow i_1 = \frac{6}{9} = 0.67 \text{ A}$$

Now substitute the value for i_1 into Equation 12:

$$0.67 - 2i_2 = 2 \Rightarrow -2i_2 = 1.33 \Rightarrow i_2 = -0.67$$

The result for i_2 is very important. Since it came out negative, your original assumption that i_2 flows clockwise is incorrect. This will work in this manner every time; regardless of what you assume, as long as you are consistent, the signs will work out at the end.

Now you can calculate the voltage drops on each resistor and then add the results appropriately to check the answer. Earlier you gave the equations of interest, so you can now calculate the results:

$$V_{R_1} = i_1 \times R_1 = 0.67 \times 24 = 16 \text{ V}$$
$$V_{R_2} = -i_2 \times R_2 = -(-.67) \times 6 = 4 \text{ V}$$
$$V_{R_3} = i_1 R_3 - i_2 R_3 = 0.67 \times 6 - (-0.67 \times 6) = 4 + 4 = 8 \text{ V}$$

Comparing these to the two loops, you see that the $24 = V_{R_1} + V_{R_3} = 16 + 8 = 24$ V and that $12 = V_{R_2} + V_{R_3} = 6 + 6 = 12$ V. Notice also that the voltage V_{R_2} came out positive, which means that although the original choice was not consistent with the assumed direction of i_2, it is consistent with the actual direction of i_2.

EXAMPLE 2

Figure 23–4 is the second example. Although it is considerably more complex, the approach is exactly the same. Table 23–1 shows the actual steps.

Solution:

Step 1

This problem requires that at least four loops be chosen. The best approach is to select the smallest number of loops possible.

Step 2

Figure 23–5 shows the four loops and the labels chosen for this problem. Note that every component is included in at least one loop.

Step 3

Figure 23–6 shows the voltage labels for each component and their polarities based on assumed direction of the loop currents.

FIGURE 23–4   Four-loop example problem for the use of Kirchhoff's voltage law.

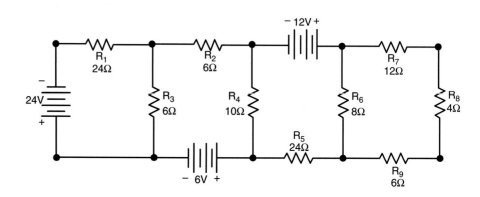

Table 23–1   Steps for the Application of Kirchhoff's Voltage Law

Step	Description	Comments
1	Using the schematic diagram, select the loops that you will use for Kirchhoff's voltage law.	You must have enough loops to ensure that all components are included in at least one loop.
2	Draw and name each loop current. You may assume any direction for the loop currents, but usually being consistent is best.	This text always uses clockwise current flow; however, you may prefer counterclockwise or even a mixture.
3	Label each component with its voltage drop.	1. Usually assume that the end of the component into which the current flows is negative. 2. If a component has current from more than one loop, you may use whichever one you wish. Usually, the first current that you consider determines polarity.
4	Write Kirchhoff's voltage law by summing all the currents around each loop and setting the sum to zero.	Starting with any component, sum its voltage (subtract if you encounter a negative sign first), then move to the next one in the direction of the assumed current flow. When you have summed all of the elements, set them to zero. Repeat this process for each loop.
5	Using Ohm's law, create a table of the values of each voltage drop.	For example, in Figure 23–3, the voltage drop $V_{R_1} = i_1 \times R_1$.
6	Substitute the Ohm's law values into the loop equations you developed in step 4.	
7	Solve the resulting equations for each loop current.	This can be done using the rules of algebra on a scientific calculator.
8	Using the currents calculated in step 7 and the table created in step 5, calculate the voltage drops for each component.	

FIGURE 23–5   Figure 23–4 with the loop currents selected.

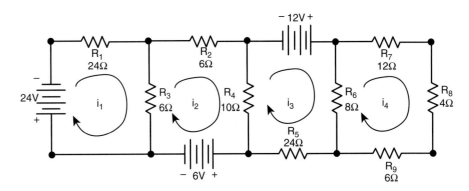

Step 4

You can now write the four loop equations starting with i_1 and progressing to i_4:

For $i_1 \rightarrow +24 - V_{R_1} - V_{R_3} = 0 \Rightarrow V_{R_1} + V_{R_3} = 24$

For $i_2 \rightarrow +6 + V_{R_3} - V_{R_2} - V_{R_4} = 0 \Rightarrow V_{R_2} - V_{R_3} + V_{R_4} = 6$

For $i_3 \rightarrow +12 - V_{R_6} - V_{R_5} + V_{R_4} = 0 \Rightarrow V_{R_4} - V_{R_5} - V_{R_6} = 12$

For $i_4 \rightarrow -V_{R_7} - V_{R_8} - V_{R_9} + V_{R_6} = 0 \Rightarrow V_{R_6} - V_{R_7} - V_{R_8} - V_{R_9} = 0$

FIGURE 23–6   Figure 23–5 with the component voltage drops labeled.

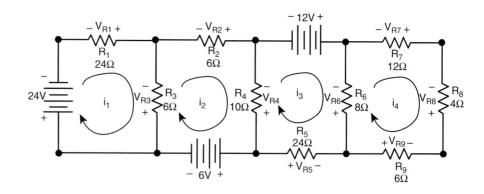

Step 5

Next, create a table of values for each voltage drop using Ohm's law:

$$V_{R_1} = i_1 R_1 = 24i_1$$
$$V_{R_2} = i_2 R_2 = 6i_2$$
$$V_{R_3} = (i_1 - i_2)R_3 = 6(i - i_2)$$
$$V_{R_4} = (i_2 - i_3)R_4 = 10(i_2 - i_3)$$
$$V_{R_5} = i_3 R_5 = 24i_3$$
$$V_{R_6} = (i_3 - i_4)R_6 = 8(i_3 - i_4)$$
$$V_{R_7} = i_4 R_7 = 12i_4$$
$$V_{R_8} = i_4 R_8 = 4i_4$$
$$V_{R_9} = i_4 R_9 = 6i_4$$

Step 6

Substituting the values into the four loop equations gives

$$24i_1 + 6(i_1 - i_2) = 24 \Rightarrow 30i_1 - 6i_2 = 24 \Rightarrow \mathbf{5i_1 - i_2 = 4}$$
$$6i_2 - 6(i_1 - i_2) + 10(i_2 - i_3) = 6 \Rightarrow \mathbf{-6i_1 + 22i_2 - 10i_3 = 6}$$
$$10(i_2 - i_3) - 24i_3 - 8(i_3 - i_4) = 12 \Rightarrow \mathbf{10i_2 - 42i_3 + 8i_4 = 12}$$
$$8(i_3 - i_4) - 12i_4 - 4i_4 - 6i_4 = 0 \Rightarrow \mathbf{8i_3 - 30i_4 = 0}$$

Step 7

Solving the equations can be done in a number of ways, but with this system it is easiest to solve them using a handheld calculator. The results are

$$i_1 = 0.885 \text{ A}$$
$$i_2 = 0.426 \text{ A}$$
$$i_3 = -0.194 \text{ A}$$
$$i_4 = -0.0518 \text{ A}$$

Step 8

The actual voltage drops can now be calculated as follows:

$$V_{R_1} = i_1 R_1 = 24i_1 = 24 \times 0.885 = 21.2 \text{ V}$$
$$V_{R_2} = i_2 R_2 = 6i_2 = 6 \times 0.426 = 2.56 \text{ V}$$

$$V_{R_3} = (i_1 - i_2)R_3 = 6(i_1 - i_2) = 6 \times (.885 - .426) = 2.754 \text{ V}$$
$$V_{R_4} = (i_2 - i_3)R_4 = 10(i_2 - i_3) = 10 \times (.426 + .194) = 6.2 \text{ V}$$
$$V_{R_5} = i_3 R_5 = 24i_3 = 24 \times -0.194 = -4.66 \text{ V}$$
$$V_{R_6} = (i_3 - i_4)R_6 = 8(i_3 - i_4) = 8 \times (-0.194 + 0.0518) = -1.14 \text{ V}$$
$$V_{R_7} = i_4 R_7 = 12i_4 = 12 \times -0.0518 = -0.622 \text{ V}$$
$$V_{R_8} = i_4 R_8 = 4i_4 = 4 \times -0.0518 = -.2072 \text{ V}$$
$$V_{R_9} = i_4 R_9 = 6i_4 = 6 \times -0.0518 = -0.311 \text{ V}$$

If you add the voltages around any loop shown in Figure 23–6, you will see that they add to zero, allowing for rounding errors.

■ SUMMARY

Kirchhoff's laws are very useful when analyzing complex circuits with more than one power source. Loop equations (Kirchhoff's voltage law) are written so that the algebraic sum of the I_R voltage drops equals zero. Care must be taken to observe the polarity and direction of current flow.

Node equations (Kirchhoff's current law) are written so that the algebraic sum of all currents entering and leaving a given circuit point must equal zero.

Remember that the point where current enters a resistive element is considered negative, and you must always be consistent with your signs after you have assumed a polarity for any value.

■ REVIEW QUESTIONS

1. Write and explain the two Kirchhoff's laws.
2. Kirchhoff's law is useful for solving more complex circuits with more than one voltage source. Discuss the advantages and disadvantages of using Kirchhoff's law versus the superposition theorem.
3. Why is it important to assign a polarity to each voltage drop in the circuit?

4. Review the steps for Kirchhoff's voltage law shown in Table 23–1. Discuss each and be certain that you understand it fully.
5. What is the assumed polarity of a resistor at the point into which the current flows?

■ PRACTICE PROBLEMS

1. Solve the circuit shown below using Kirchhoff's voltage law.

2. Solve the circuit again using the superposition theorem.

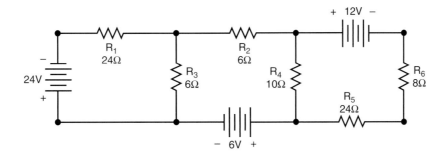

Thevenin's and Norton's Theorems

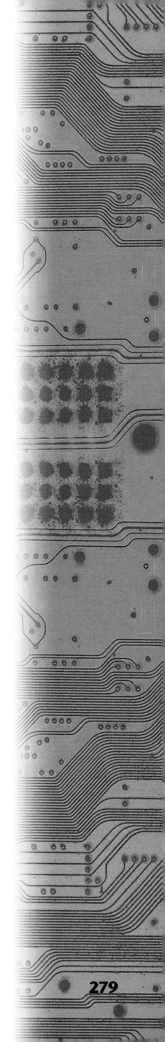

■ OVERVIEW

Thevenin's and Norton's theorems can save you many hours recalculating entire circuits for changing loads. They are used extensively in electronic circuit applications and also have application in many electrical power system applications. They can be especially useful when dealing with many unknowns in complex circuits, such as those encountered during network analysis in a power system. In this chapter, you will learn about both of the theorems and how to use them.

■ OBJECTIVES

After completing this chapter, you should be able to:

1. Explain Thevenin's and Norton's theorems.
2. Apply Thevenin's and Norton's theorems to solve for circuit unknowns.
3. View circuits in terms of open-circuit voltages and maximum circuit currents.

FIGURE 24–1   The principle of the Thevenin equivalent circuit.

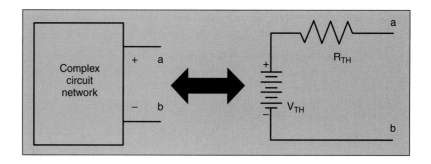

■ THEVENIN'S THEOREM

24.1 Introduction

A French engineer, M. L.Thevenin, invented this theorem to simplify voltages in a network. No matter how many sources and components and no matter how they are interconnected, with Thevenin's theorem they can all be represented by an equivalent series circuit with respect to any "pair" of terminals in the network (see Figure 24–1).

Look at Figure 24–1. Imagine that the block on the left contains a network connected to terminals a and b. Thevenin's theorem states that the entire network connected to a and b can be replaced by a single voltage source (V_{TH}) in series with a single resistance (R_{TH}) connected to the same terminals. The definitions of these two values are as follows:

1. V_{TH} is the open-circuit voltage measured at terminals a and b.
2. R_{TH} is open-circuit voltage (V_{TH}) divided by the short-circuit current (I_{SC}) that would occur between terminals a and b.

The following sections illustrate methods that can be used to determine and check a Thevenin equivalent circuit.

24.2 Creating a Thevenin Equivalent Circuit

Overview

You wish to create a Thevenin equivalent circuit for terminals a and b of Figure 24–2. This means that any load that is attached to terminals a and b on the Thevenin circuit will result in the same voltage and current as if the same load were attached to the original circuit. The following explanation shows how to find the Thevenin equivalent values.

Finding V_{TH}

V_{TH} is the open-circuit voltage at terminals a and b. This is clearly equal to the voltage drop across resistor R_2, which can be calculated using the voltage divider formula as follows:

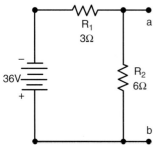

FIGURE 24–2   Sample circuit for creating a Thevenin equivalent circuit.

$$V_{\text{TH}} = V_{R_2} = 36 \times \left(\frac{R_2}{R_1 + R_2}\right) = 36 \times \frac{6}{3 + 6} = 24 \text{ V}$$

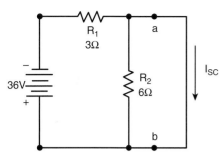

FIGURE 24–3 Calculating I_{SC} for Figure 24–2.

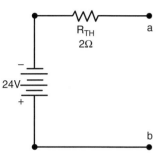

FIGURE 24–4 Thevenin equivalent circuit for Figure 24–2.

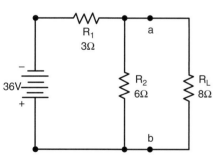

FIGURE 24–5 Figure 24–2 with an 8-ohm load connected.

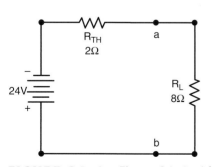

FIGURE 24–6 Figure 24–4 with an 8-ohm load connected.

Finding R_TH

R_{TH} can be found by first placing a short circuit at points a and b and then measuring or calculating the current that flows between a and b. From Figure 24–3, I_{SC} can be calculated as

$$I_{SC} = \frac{36}{3} = 12 \text{ A}$$

R_{TH} is now calculated as follows:

$$R_{TH} = \frac{E_{TH}}{I_{SC}} = \frac{24}{12} = 2 \text{ }\Omega$$

Note that R_{TH} can also be calculated by turning the voltage sources to zero (short circuit) and then calculating the equivalent resistance between terminals a and b. The complete Thevenin equivalent circuit is shown in Figure 24–4.

Checking Results

If your calculations have been accurate, any load placed at terminals a and b in Figure 24–2 should create the same load voltage (V_L) and load current (I_L). First check the original circuit with an 8-ohm load connected as shown in Figure 24–5.

The voltage across the load can be calculated by first realizing that R_L and R_2 are connected in parallel. The equivalent resistance of these two is

$$R_{L,2} = \frac{R_L \times R_2}{R_L + R_2} = \frac{6 \times 8}{6 + 8} = 3.43 \text{ }\Omega$$

Now the voltage drop across R_L can be calculated by using the voltage divider formula:

$$V_{L,2} = 36 \times \frac{R_{L,2}}{R_{L,2} + R_1} = 36 \times \frac{3.43}{6.43} = 19.2 \text{ V}$$

Since the voltage across $V_{L,2}$ is the same as the voltage across V_L, then $V_L = 19.2$ V.

Next check the voltage drop across an 8-ohm load connected to terminals a and b of the Thevenin equivalent circuit as shown in Figure 24–6. The simple approach here is to use the voltage divider again:

$$V_L = 24 \times \frac{8}{8 + 10} = 19.2 \text{ V}$$

Clearly, the Thevenin equivalent circuit of Figure 24–4 performs exactly as the original circuit in Figure 24–2.

■ NORTON'S THEOREM

24.3 Introduction

An American, E. L. Norton, who worked for Bell Telephone Laboratories, invented this theorem. This theorem simplifies complex circuit networks using current sources instead of voltage. This circuit replaces the Thevenin circuit voltage source and series resistor with a current source and parallel resistor. Figure 24–7 shows the comparison between Thevenin and Norton circuits.

The concept is that the load you place on terminals a and b for any one of the three circuits will create the same voltage and current. The two quantities needed are defined as follows:

1. I_N is the measured current between terminals a and b when a short circuit is applied.
2. R_N is calculated the same as R_{TH}. It is the open-circuit voltage (V_{ab}) divided by the short-circuit current (I_{SC}) that would occur between terminals a and b. As with Thevenin's theorem, it can also be calculated by turning all sources to zero and calculating the resistance between terminals a and b.

24.4 Why a Current Source?

The Norton circuit is especially useful when analyzing circuits that have parallel resistances. The series resistance equivalent of a Thevenin circuit would register varying voltages on parallel loads based on the different current through the resistances. Norton's theorem allows a constant current across the varying loads and thus easier voltage calculations.

24.5 Creating a Norton Equivalent Circuit

Overview

You wish to create a Norton equivalent circuit for terminals a and b of Figure 24–2. This means that any load that is attached to terminals a and b on the Norton circuit will result in the same voltage and current as if the same load were attached to the original circuit. The following explanation shows how to find the Norton equivalent values.

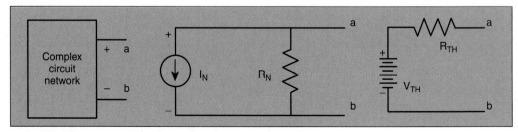

FIGURE 24–7   Thevenin and Norton comparison.

Finding I_N

I_N is the current between terminals a and b when they are short-circuited. This was calculated earlier as

$$I_{SC} = \frac{36}{3} = 12 \text{ A} = I_N$$

The circuit that was used to develop this value is shown in Figure 24–3.

Finding R_N

Calculate R_N as before or by turning the voltage source in Figure 24–2 to zero and calculating the impedance at terminals a and b. Figure 24–8 shows the method.

R_{TH} is now calculated as follows:

$$R_N = \frac{R_1 \times R_2}{R_1 + R_2} = \frac{18}{9} = 2 \text{ }\Omega$$

Note that R_N can also be calculated by turning the voltage sources to zero (short circuit) and then calculating the equivalent resistance between terminals a and b. The complete Thevenin equivalent circuit is shown in Figure 24–9.

Checking Results

If your calculations have been accurate, any load placed at terminals a and b in Figure 24–2 should create the same load voltage (V_L) and load current (I_L). Earlier, you calculated the voltage drop with an 8-ohm load connected to the circuit (Figure 24–4).

FIGURE 24–8   Calculating R_N.

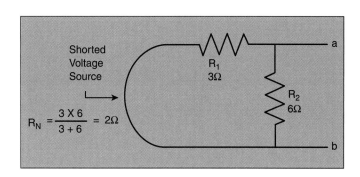

FIGURE 24–9   Norton equivalent circuit for Figure 24–2.

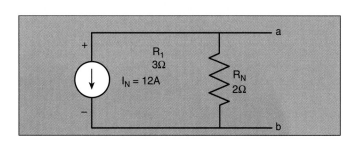

FIGURE 24–10 Figure 24–9 with an 8-ohm load connected.

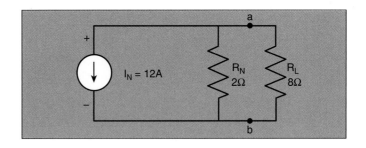

The voltage across the load can be calculated by realizing that R_L and R_2 are connected in parallel. The equivalent resistance of these two is

$$R_{L,2} = \frac{R_L \times R_2}{R_L + R_2} = \frac{6 \times 8}{6 + 8} = 3.43 \ \Omega$$

Now the voltage drop across R_L can be calculated by using the voltage divider formula:

$$V_{L,2} = 36 \times \frac{R_{L,2}}{R_{L,2} + R_1} = 36 \times \frac{3.43}{6.43} = 19.2 \ \text{V}$$

Since the voltage across $V_{L,2}$ is the same as the voltage across V_L, then $V_L = 19.2$ V.

Next, check the voltage drop across an 8-ohm load connected to terminals a and b of the Norton equivalent circuit as shown in Figure 24–10. You can start by using a current divider formula to calculate the current through the load resistor:

$$I_L = 12 \times \frac{R_N}{R_L + R_N} = 12 \times \frac{2}{10} = \frac{12}{5} \ \text{A}$$

$$V_{RL} = 8 \times \frac{12}{5} = 19.2 \ \text{V}$$

Clearly, the Norton equivalent circuit of Figure 24–9 performs exactly as the original circuit in Figure 24–2 and the Thevenin equivalent in Figure 24–9.

■ SUMMARY

As discussed in this chapter, Thevenin's theorem says that any network can be represented by a voltage source and series resistance, while Norton's theorem says that the same network can be represented by a current source and parallel resistance. In fact, the same example was used for both theorems, and with the same "new" load applied to both types of equivalent circuit, the resulting voltage drops were the same: 19.2 volts. These two theorems will prove useful in circuit analysis for both power systems and electronic circuits.

■ REVIEW QUESTIONS

1. Where are Thevenin's and Norton's theorems generally used?

2. Why did Thevenin and Norton invent their theorems?

3. State and discuss Thevenin's theorem.

4. State and discuss Norton's theorem.

■ PRACTICE PROBLEM

1. For the following circuit,

  a. Determine the Thevenin equivalent circuit at terminals a and b.

  b. Determine the Norton equivalent circuit at terminals a and b.

  c. Prove your answer by attaching a 10-ohm resistor to each circuit (at terminals a and b) and show that the same voltage and currents are present.

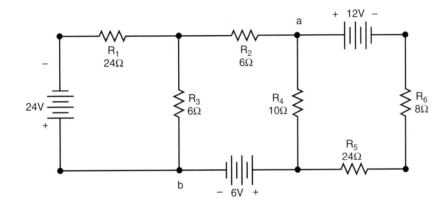

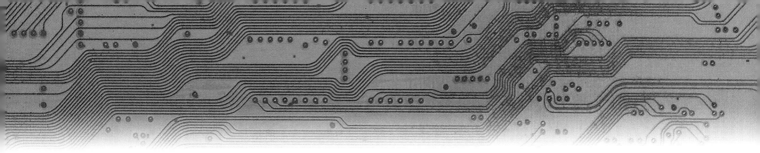

Index